D0934244

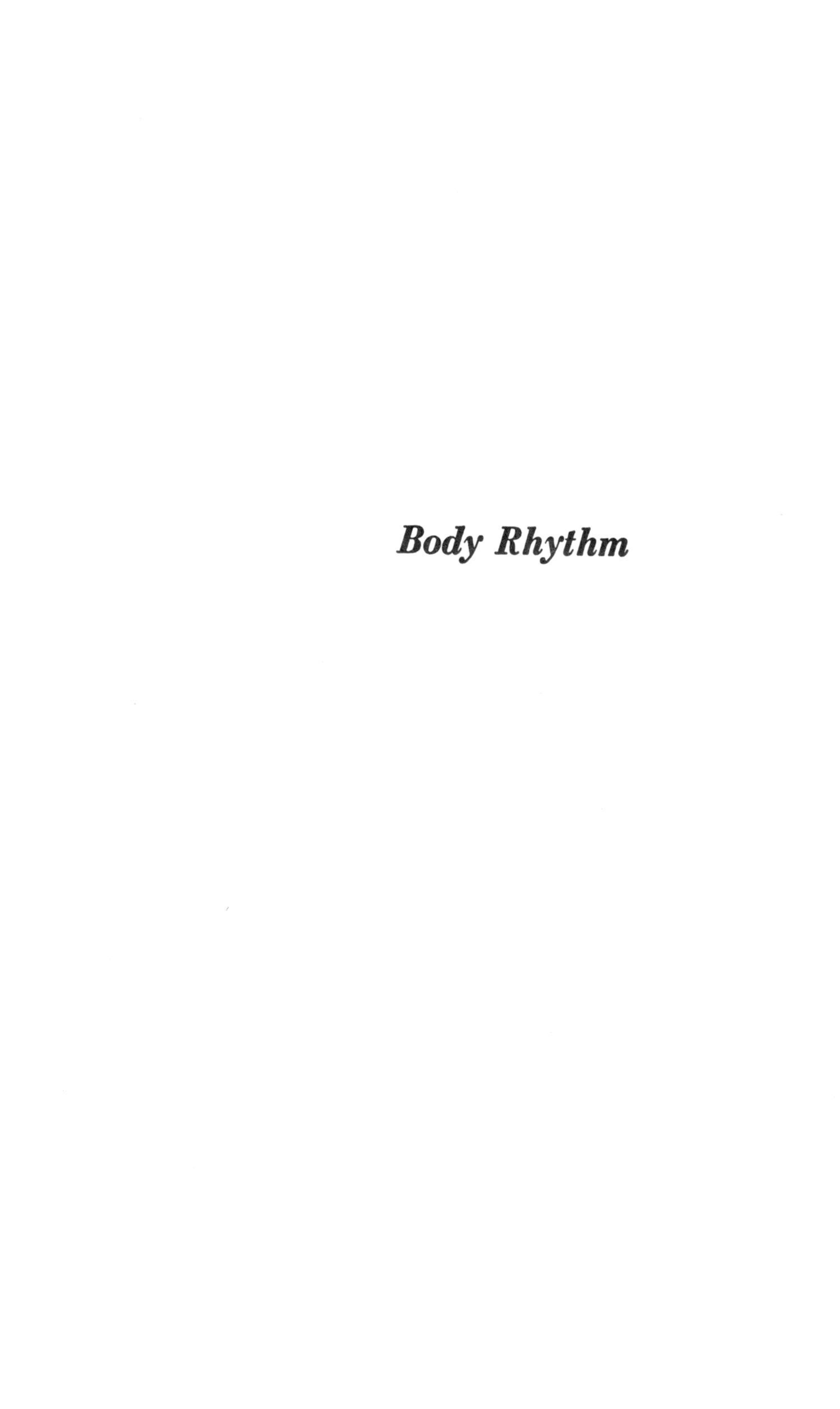

Body Rhythm

Lee Weston

Body Rhythm

The Circadian Rhythms Within You

Harcourt Brace Jovanovich

New York and London

Printed in the United States of America

Library of Congress Cataloging in Publication Data

*Weston, Lee.
Body rhythm: the circadian rhythms within you.*

*Bibliography: p.
Includes index.
1. Circadian rhythms. 2. Biological rhythms.
I. Title.
QP84.6.W47 612 79–1854*
ISBN *0–15–113338–7*

First edition

B C D E

Acknowledgments

The author wishes to express his grateful appreciation to Colin Pittendrigh of Stanford University, whose lecture inaugurated this project and who was gracious enough to include me in the sessions of his workshop on rhythms in mammals and man; to Charles Winget of NASA, who brought awareness of the scope of the field; and to all the others who so generously shared their time and knowledge: Jürgen Aschoff, Director, Max Planck Institute for Physiology of Behavior; William C. Dement, School of Medicine, Stanford University; Bruce J. Durie, University of Edinburgh; Charles F. Ehret, Argonne National Laboratory; John D. Fernstrom, Massachusetts Institute of Technology; Daniel F. Kripke, Department of Psychiatry, University of California, San Diego; Franz Halberg, University of Minnesota Medical School; Cheryl Hart, School of Medicine, University of Washington; Stanley R. Mohler, Office of Aviation Medicine, Federal Aviation Administration; Martin C. Moore-Ede, Harvard Medical School; Don A. Rockwell, School of Medicine, University of California, Davis; Ray Rosenman, Harold Brunn Institute, Mt. Zion Medical Center, San Francisco; Gary Richardson, Montefiore Medical Center, New York; Laurence E. Scheving, College of Medicine, University of Arkansas; Hugh W. Simpson, Royal Infirmary, Glasgow; Elliot D. Weitzman, Albert Einstein College of Medicine, New York; Joan Vernikos-Danellis, NASA/Ames; and the many in the field of aviation who were so helpful.

Contents

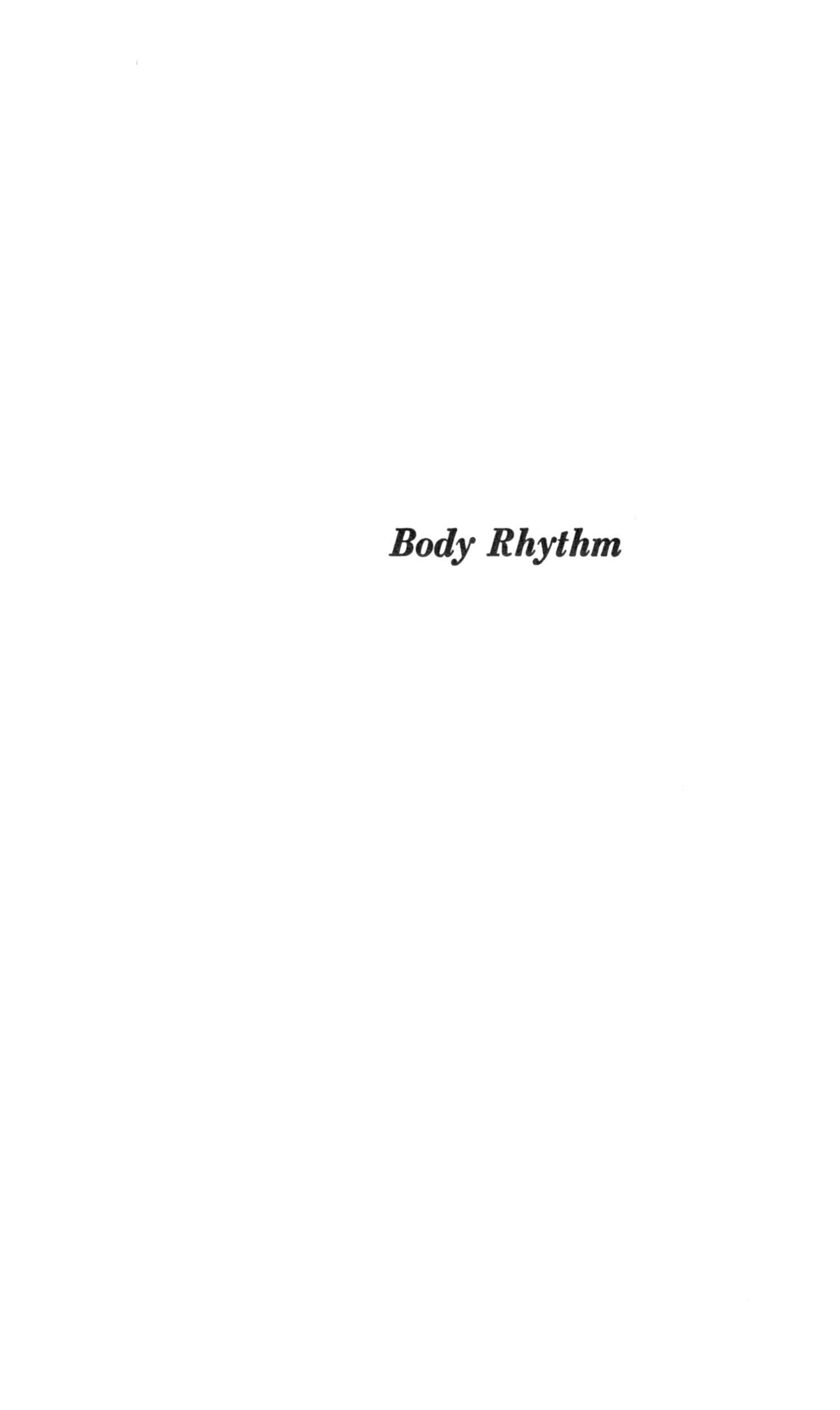

Body Rhythm

The hidden harmony is better than the obvious.

—Heraclitus

Introduction

Suppose, for a moment, that the human body is a symphony orchestra, composed of many distinct instruments that have distinct capabilities and responsibilities. The heart is the percussion section, providing a tempo. The brass section, which blares loudly at times but is muted or at rest at others, resembles the adrenal glands. The respiratory system is ceaseless, complete with crescendos and pianissimos, functioning as the strings do in a serenade. And the brain is the conductor, instructing the players of their duties, keeping them in line.

Dozens of other instruments contribute to the harmony of the whole, but one more element bears particular mention. The score, for an orchestra, stands between cacophony and beautiful music. Without it, the musicians produce noise—jarring, disturbing sounds. The score for the human body is its circadian rhythms.

The term *circadian* comes from the Latin *circa,* meaning "about," and *dies,* meaning "day," and was coined by Dr. Franz Halberg of the University of Minnesota. Circadian rhythms are the body's inherent daily rhythms pulsing through each of us every day. In combination with a "clock," or a mechanism that acts like a clock, these rhythms keep an individual in phase with his environment. The rhythms repeat themselves every day throughout the life of the organism (be it man or a lower animal, or even a plant), marking the passage of time like a metronome.

No one organ is responsible for the body's circadian rhythms. They are a composite of the organism's functions, its many processes, and their ebbs and flows. To illustrate the cycles of these rhythms, let us look at a typical person's circadian cycle.

George (or Georgette if you prefer—the male and female rhythms are seldom different) Average is, as his name suggests, average. He follows a similar routine almost every day. He wakes at about the same time each morning, often without the aid of the insistent and irritating buzz of his alarm clock. As his mind gradually returns from dreamland, he is reluctant to leap up from his cozy bed—his body temperature is near its nadir. He may be more inclined to turn to his wife and propose that they make love. She might well agree: both the male and the female hormones flow more quickly in the early hours of the morning as the body temperature gradually rises.

It is Saturday, and George is not limited to the routines of the office or the shop but can follow his own inclinations. When he gets up, his mood is likely to be serene, and he will not relish intrusions. He may use the barrier of a newspaper or the droning of the radio to drown them out. That his breakfast is bland is appropriate; his sensory perceptions, including taste, are low.

If he's been lax in his correspondence of late, he might well sit down and catch up at this point in the day. His memory works best in the morning but trails off around noon.

As the morning passes, he feels increasingly energetic. At about 11:00 A.M. his body temperature reaches a plateau where it remains with only slight variation, until shortly before he returns to bed for the night. If he has to balance his checkbook, he would be well advised to do it just before lunch, as his mental skills, such as speed and accuracy in multiplication, reach their high point around midday and decline during the afternoon.

George will probably experience an after-lunch letdown. His performance levels will dip temporarily, and he may feel inclined to take a short nap. His temperature may even fall slightly just after lunch.

Let's say his lawn mower broke last week. His project for the afternoon is to fix it. Good timing: George's mechanical skills rise to their peak in the afternoon. His motor coordination stays level most of the day, so he is also comfortable with his screwdriver and pliers.

After the mower is fixed and his lawn is as trim and green as his neighbor's yard, George probably feels more cheerful and gregarious than he has all day. As he walks into the kitchen, the stew that simmers on the stove may smell even better than it really is. The neighbor's kids, irritating little brats that they are, seem even more so.

His senses are now at their peak. Dinner is a delight because his taste buds are at their peak sensitivity, but Junior's guitar practice is a nightmare. As his body winds down, his temperature begins to drop, especially after 10:00 P.M. His evening cocktail has the desired relaxing effect, but it doesn't make him as tipsy as it would have if he'd had it in the morning.

When he goes to bed, he slips almost immediately into a deep or "Delta" sleep. As the night wears on, he spends less time dead to the world, his mind inactive, and more time in the stage of sleep known as REM, or Rapid Eye Movement, when he dreams. His temperature falls throughout the night until about 5:00 A.M., when it reaches its low point. An hour or so before that, a variety of hormones are secreted by George's glands, preparing the metabolism of his body and mind to gear itself for another day. As they have their effect, his body temperature begins to rise from its 5:00 A.M. low point.

George is an average man, but as with all averages, he is typical only insofar as he is the product of other people's differences. Some people wake up alert and ready to do battle; some are somnolent until most people are ready to relax, when they come to life. Whatever the variations from individual to individual, circadian rhythms are constants in each and every healthy person's life. Mental and physical illnesses can disrupt the body's synchronization, but a healthy person has his own constant rhythm.

In the following chapters we will look at the workings of the human body while it is awake and while it sleeps, in sickness and in health. There is much to be learned about ourselves and about an area of biology that, since its sudden emergence in the past twenty-five years, has come to be regarded as a characteristic of virtually all forms of life. This book is both an introduction to the rhythms of life discovered by circadian science and a guide to the workings of our bodies and minds with respect to their rhythms, which control our ability to perform as well as enjoy.

Scientists are only beginning to study, report on, and comprehend the daily cycles of George Average, not to mention the confused functioning of all the George Aberrants who inhabit this earth. But let us now travel to the beginning, where man's knowledge of this orchestration began. And as we travel through time, it will become clear how far we have come, but also how much farther we have yet to go.

The advance of true natural philosophy, which is experimental, can only be painfully slow.

—Marchant (writing for Jean Jacques d'Ortous de Mairan)

1 *The Rhythms of Nature*

The goddess Athena came into the world in a highly unusual manner: she was born from her father's forehead. As if her means of entry into the world was not remarkable enough, her form, too, must have raised a few eyebrows. At birth she was clad in armor, a mature woman and warrior.

Her birth is the stuff of which ancient myth is made, but it can stand as a metaphorical contrast to the study of circadian rhythms. Knowledge of circadian rhythms has been obtained by gathering bits and pieces of theory and data by fits and starts, information that has been lying dormant for many decades between bursts of activity. More often than not, man discovers what is in front of him only when he trips over it, and the first observation of what we now call circadian rhythms came as the result of plain serendipity.

A French astronomer named Jean Jacques d'Ortous de Mairan labored mightily on a variety of projects. (He, like many eighteenth-century scientists, was a generalist, a "natural philosopher.") Between bouts of concentrated study of eclipses and the geology of the Alps, de Mairan observed that a plant in his study opened its leaves each morning and closed them every night. He also noticed that the plant, as if on command, folded and unfolded its appendages at the same time each day.

He decided to try to trick the plant, a heliotrope, in order to discover what caused its peculiar behavior. Perhaps it was reacting to light, he hypothesized, exposing its leaves to the light of day, then closing upon itself at night. He placed the plant in a darkened room. The leaves continued to open and close, keeping to their appointed schedule, even without the changes from light to dark and back again.

Eventually, de Mairan gave up and returned to his other interests. He refused even to write a paper on the phenomenon he had discovered. It was left to a colleague, as was often done in the early days of science, to present the findings to the Royal Academy of Paris.[1]

That was in 1729. Little theory was added to de Mairan's work for almost two centuries, though some further observations were recorded. The Swedish naturalist Carolus Linnaeus (who devised a system of plant classification in the mid-eighteenth century) discovered that a number of flowers opened and closed their petals at certain times of day. His observations had more of an effect on the castle-and-courtyard set than on serious scientists, however, as gardens were constructed using his findings to delight the eye rather than to engage the inquisitive mind.

In 1758 Henri-Louis Duhamel duplicated de Mairan's experiments. He went on to demonstrate that the opening and closing of the leaves was independent of temperature. When he set some sensitive plants in his temperature-controlled hothouse, nothing changed. The heliotrope leaves continued to flex and unflex on schedule.[2]

After the turn of the nineteenth century, a Swiss botanist named Augustin Pyramus de Candolle conducted a series of experiments with the *Mimosa pudica,* a plant that also manifested daily leaf changes. After confirming de Mairan's and Duhamel's experiments, he added another variable. He isolated his mimosa from sunlight and directed the beams of six lamps onto it instead. The intensity of this artificial illumination was approximately that of a sunless day.

He never shut the lights off but left them lit twenty-four hours a day. The plant remained unperturbed, continuing its cycle of "sleeping" and "waking." But one factor changed: the mimosa shortened its cycle and traveled its entire schedule of opening, closing, and reopening in twenty-two to twenty-two and a half hours.[3] This phenomenon was little understood then (and was probably perplexing indeed), but it is referred to now as "free running." As we shall see later, free-running rhythms are not only common but are characteristic of organisms when light and other time cues are removed.

The seminal genius Charles Darwin contributed to circadian rhythm also, as he experimented with the effects of light on plants (his treatise was entitled *The Power of Movement in Plants*); and Wilhelm Pfeffer, the so-called father of plant physiology, experimented with the effects of gravitational force on plant movements. But it was not until World War I that practical research began.[4]

Two researchers at the Department of Agriculture, H. A. Allard and W. W. Garner, set out to solve the conundrum of the Maryland mammoth. An enormous tobacco plant, the mammoth bore many fine leaves but in the area around Washington, D.C., it fell prey to frost before it could flower and produce seeds.

Allard and Garner had noticed that several plots of soybeans, planted at intervals, all produced their blooms at the same time. This observation helped them solve the problem of the seedless mammoth and to conclude that the mammoth ". . . merely inherits the capacity to flower and fruit in response to certain favorable external conditions."[5]

After two years of literally fruitless research, they built a light-proof, ventilated "dog house" (their term) and placed pots of Maryland mammoth in it for fourteen hours a day. They exposed the plant to ten hours of daylight, comparable to the length of daylight during the fall—although it was actually the frostless month of July. Under the ten hours' sunlight, the Maryland mammoth bloomed. Allard and Garner's discovery that the tobacco plant responds to the cycles of the sun proved to be the curtain raiser for future discoveries of circadian rhythms and clocks.

Farmers contemporary to Allard and Garner found that hens would lay more eggs if the lights were left on longer. The Japanese and Dutch discovered that artificially longer days kept birds singing longer.[6] The German biologist Erwin Bunning devoted his life to exploring the internal clocks in plants and fruit flies. But only when Karl von Frisch and Gustav Kramer began their researches was the study of circadian rhythms truly inaugurated.

Their research marks a watershed. Kramer studied starlings and von Frisch bees. Both animals were found to use an internal clock that compensated for the movement of the sun to direct them to their summer or winter homes or in the direction of flowers. Reports began flowing in from a variety of sources around the world, demonstrating the presence of circadian rhythms in numerous organisms and even indicating their involvement in the secretion of some hormones in man.

In an inspired moment of synthesis, Colin Pittendrigh, then of Princeton University and now at Stanford, recognized from his own work and that of others—from de Mairan to Kramer and von Frisch—the inevitable hypothesis that circadian rhythms are a virtually ubiquitous characteristic of life. All that remained to establish circadian rhythms firmly as an independent area of research was Franz Halberg's coinage of the term *circadian.* From the floundering incomprehension of de Mairan, it had taken better than two centuries, but in 1954 a field of science was founded.

The study of circadian rhythms falls into a broader category, the science of *chronobiology.* As its name indicates, chronobiology is the study of living organisms with respect to time. Circadian rhythms lie at the heart of chronobiology—the twenty-four-hour unit of one solar day almost certainly cues the other changes, whether they are monthly, seasonal, annual, or of any other duration.

Nature's rhythms can be of virtually any length. The rapid rhythms of our brain last but a split second. At the other end of the spectrum, nature's longest rhythm is perhaps that of the *Phyllostachys bambusoides,* a species of bamboo native to China. The first reported flowering of the species was in A.D. 191, and since then it has flowered at 120-year intervals with unerring regularity, regardless of its location. In the late 1960s it flowered simultaneously in Japan, England, Russia, and Alabama.[7]

In fact, most of the hundred-plus varieties of bamboo have a flowering cycle of fifteen years or more, wherever they grow (except at the Equator), and whatever their height—six inches or forty feet—and they will flower at their appointed time, simultaneously. There is no known environmental cycle to account for the synchronous flowerings, and bamboo must be regulated by an internal, inherited "clock."

Thus we return, inevitably, to circadian rhythms. But somewhere between the cycles of the brain and the flowering of the bamboo lies a pseudoscience that is often confused with circadian rhythms.

Biorhythms made their debut almost a century ago, but only in recent years have they truly come into vogue. In a television program prior to the 1978 Super Bowl, a trio of sports announcers revealed to a national audience that the biorhythm charts of the Denver Bronco players were superior to those of the Dallas Cowboys. The Cowboys thereupon beat the Broncos, biorhythms or not.

In 1887 Wilhelm Fliess concocted the theory that lurks behind biorhythms. He postulated that everyone is bisexual, possessed of a male component of strength, endurance, and courage and a female component of sensitivity, intuition, and love. The cycles are twenty-three and twenty-eight days in duration, respectively. These elements, according to Fliess, are to be found in every cell of the body, and particularly in the nose.[8]

In the proboscis certain conspicuous "genital" cells reside, said Fliess, and he claimed to be able to diagnose neurotic symptoms and sexual abnormalities from them. He treated them by surgery or by the application of cocaine. Freud proclaimed Fliess's discovery to be "a great scientific breakthrough" and actually had two nasal operations himself. Had Fliess not been allied to Freud, perhaps his pet theory would have died mercifully in its own time.[9]

Instead, biorhythms lived on. In the 1930s a final refinement was added to the system. It was a third cycle, one of mental acumen and power, thirty-three days long. That completed the triad of body, mind, and emotions.

According to the theory, the biorhythms begin on the day of one's birth, manifesting themselves in life's ups and downs and eventually dictating the day of death. The elements are all positive when one is born, and as we go through life they deliver us the good times and bad, depending upon their constant movement and how they are juxtaposed to one another.

Since biorhythm charts are easy to construct, the comparison to reality is simple. From an examination of the records of 8,625 pilots involved in aviation mishaps from the files of the army, navy, Federal Aviation Administration, and National Transportation Safety Board, negative biorhythms were flatly ruled out as a cause of the accidents.[10] In 1971 the Workmen's Compensation Board of British Columbia conducted a study of more than 13,000 cases. The research statistician of the board concluded, "The results indicate that accidents are no more likely to occur during so-called critical periods than at any other time."[11]

Thus, many researchers share Franz Halberg's sentiment: "No serious scientist has ever supposed for a moment that there was any foundation to the biorhythms theory." But the process of definition can be clarified by discussion of what something is not. We can now clear the air—biorhythms most certainly do not play any role in the serious study of chronobiology.

So let's now flip the coin and return to looking at what circadian rhythms are. When Mother and Father, according to the sociological myth, sat us down to teach us about sex, they started with the birds and the bees. And in postwar Germany, Herr von Frisch was working with bees and Herr Kramer with birds.

There opened before us a view of the unexpected new territory.

—Karl von Frisch

2 Internal Clocks and the Animal Kingdom

As any experienced picnicker knows, the presence of the food-filled picnic basket guarantees the arrival of at least one uninvited guest: the ant. If you feel like being vengeful (or merely inquisitive), you can play a simple trick on the segmented insects sharing your food.

As the battalion of ants that have just pillaged your otherwise enjoyable lunch on the lawn heads home, shield them from the direct rays of the sun. Then use a mirror to reflect the sunlight back onto them, but from a different angle. The ants will turn, and their new direction will maintain the same angle to your "sun" as their original line of march had to the real sun.

Ants use the sun to find their bearings (you will be directing them to points unknown). The insects compensate for the changing position of the sun by using an internal clock; they know where the sun will be at a certain hour, so they travel at a certain angle to its rays to return to their communities.[1]

Although this experiment was first conducted in 1911, the

notion that such simple organisms—or any organisms at all—had such a clock did not fit into the thinking of the times. The knowledge lay fallow for forty years, until Kramer and von Frisch began their research.

Like many laymen and scientists before him, Karl von Frisch noticed that bees could be lured to a particular place at a particular time each day merely by offering them sweets. With that observation as a point of origin, von Frisch conducted a number of experiments. In one he captured the bees at the feeding station (they had arrived, of course, right on schedule) and secluded them in total darkness for two hours. When released, the bees flew unerringly back to their hives. Von Frisch concluded that even while the bees were detained, their internal clocks kept running and they were able to compensate for the changed position of the sun when they emerged.[2]

Through the work of von Frisch and others, we now know that bees communicate information about food sources to their brethren. The bee performs a carefully choreographed "dance" to convey the direction of and approximate distance to a food source. The "waggle dance," for example, communicates information about long-distance dinners, and the slower the waggle, the farther the food. Even if the bee discovers its supper at dusk, it will waggle all night and the axis of the dance will rotate 15 degrees each hour (the same speed as the rotation of the earth). Clearly, the bee must have some time mechanism which allows it to maintain the pattern without the sun's cues.[3]

The bee can also communicate a path it didn't travel. A bee's antennae and the hair on its eyes sense crosswinds, and the insect compensates for the effects of the breeze. When a bee returns to the hive from a nectar source, having battled a crosswind, it will always dance the angle to the sun, not the actual path it flew.[4] It can do so only by having a very accurate clock and sense of time.

Birds, too, use the sun to judge where they are, but in a more complex fashion. The bird exhibits true navigation, an exact knowledge of its location, while the bee merely uses its sense of time in combination with the sun to select the proper direction.[5]

Starlings that breed in the Baltic region fly southwest through the Low Countries to their winter quarters in southern England and Ireland.[6] If adult birds are captured and taken to Switzerland, upon their release they fly northwest rather than southwest to find their usual winter havens. If a bee is so displaced, it is geographically disoriented and continues flying southwest, ending up in southern France.[7]

In short, the navigational systems on which man spends millions of dollars are present in nature in a bird's brain. Man seems to imitate the bird in another way as well: the life-style our feathered friends maintain of spending the winter in the south is envied, and often imitated, by man.

But migration is actually just one part of a larger scheme of mating and nesting, disease prevention, genetic control, and evolution. The call to fly elsewhere has a number of manifestations in the bird's metabolism. Its reproductive functions become dormant, and it sings less (there is no need to be adamant about a territory soon to be abandoned). Like an animal that hibernates, the bird begins to store fat to fuel its flight.

Migratory birds caged in the spring and fall exhibit migratory flutters, as Gustav Kramer, of the Max Planck Institute for Marine Biology in Wilhelmshaven, Germany, discovered. The birds become quite agitated and stand in the corner of the cage pointing toward the direction they should be flying.

Kramer built an elaborate cage with shutters and mirrors with which he could test the effects of shifting the direction of the sunlight. If a starling was fluttering in the southwest corner of the cage at noon, Kramer would shift the light 90 degrees so that it was coming from the west rather than from the south. The bird, Kramer found, would turn 90 degrees, maintaining the same angle to the sun, and flutter to the northwest.[8]

Kramer performed more complicated experiments, taking great care to camouflage any clues that might tip off the bird. He conducted his experiments in a circular cage that eliminated any directional clues the bird might derive from a less regularly shaped enclosure. In one experiment he trained a bird to eat from one of twelve identical feeding stations. Eleven were empty, but the twelfth, located in the western sector of the cage, contained seed. After a month of eating at dawn, the bird was acclimated to eating at the feeder opposite the rays of the sun.

One day Kramer placed food in the east feeding station in the late afternoon, when the sun was coming from the opposite direction. Remarkably, the bird knew that it should reverse its direction almost 180 degrees with respect to the sun—it was the opposite end of the day, after all—and the starling flew to its dinner.[9]

The ultimate proof of a bird's inner clock doesn't come from within the cage but from without. Depending upon the season, the duration of the sun's light increases or decreases several minutes each day. If a bird did not have an accurate clock allowing it to compensate for the changing of the sun's arc, its navigation would be askew, and its flight path would become increasingly erratic. But the bird's system is supremely accurate. The slender-billed shearwater nests in Australia in November, then flies up the coast of Asia and across the Aleutian Islands, down the California coast, and then back across the Pacific to Australia. After all that, banded birds have returned to Australia to lay eggs on the same day for four consecutive years.[10]

The trigger for this remarkable navigational system is *photoperiodism.* The term was coined back in 1920 by H. A. Allard and W. W. Garner in their paper about the Maryland mammoth tobacco plant. Photoperiodism, said Allard and Garner, is the ability of living organisms to anticipate the shift in seasons by sensing the length, or period, of daylight. A plant or animal makes the appropriate adjustments in itself to cope with upcoming events.[11]

Since their discoveries, it has been established that the period of darkness, rather than light, triggers the reaction, but the term *photoperiodism* has stuck. Many data have reinforced the notion that changes in the composition of the light-dark cycle cause the organism to change its physiology to bloom, shed, molt, migrate, mate, and hibernate.

Photoperiodism is a seasonal strategy developed through natural selection to enable the organism to cope with the challenges brought by each changing season. Where climatic challenges are absent, the phenomenon is weak or virtually absent. The wide-awake tern breeds every 9.6 months and the brown booby is on an 8-month schedule. They are tropical seabirds; the wonder is not that they lack an annual rhythm fostered by photoperiodism but that nature still imposes a well-marked synchrony instead of permitting random or continuous breeding.[12]

Photoperiodism is one of the "entraining" processes Mother Nature requires of the denizens of her world. The purpose of photoperiodism is simply survival: an organism's inherent circadian rhythms are synchronized with those of its environment so that it will behave properly at any given season over the course of the year.[13] Many of our circadian rhythms have annual and seasonal fluctuations, and photoperiodism influences these longer, rhythmic fluctuations.

Our daily rhythms also have a survival function. It is now thought that the rat became a nocturnal animal in response to daytime predators. Animals such as the rabbit, which are the constant prey of others, have more than one rest-activity cycle within their daily rhythm, since they cannot afford the luxury of a long period of sleep. Even man has some vestiges of a protective rhythm. Our senses are at their peak during the late afternoon and early evening, ready to alert us to possible predators until the protection of night envelops us. It may seem an anachronism today, but to our forefathers huddling in their caves at twilight it was a very practical (if unrecognized) rhythm.

Being entrained to the temporal environment permits an animal to find food when it is available. The petals of plants, as well as the leaves, have circadian rhythms. The bee's rhythms not only direct it to the nectar but also see to it that, for example, it will be in the buckwheat field only in the morning, since the buckwheat blooms secrete their nectar only early in the day.[14]

Man is not quite as reasonable as nature. If we were, we wouldn't be forever nattering on about Indian summers or how it always seems to rain on the weekend. Were the organisms around us to rely on temperature or humidity for their time cues, they would be hopelessly confused. Nature is cued to light, since light is "noise free," without false or confusing information.[15]

Nature leaves nothing to chance, but makes allowances for the unexpected as well as the usual. The hamster can rely on a circannual rhythm to initiate its mating season, as well as photoperiodism. As the days lengthen in the spring, the hamster comes into season, and as shorter days arrive in the fall, its mating instinct bows out. But if the hamster is shielded from its accustomed light cues in a laboratory, it nevertheless comes into the mating mode twenty-five weeks after its retreat in the fall. A circannual signal in its clock assures the preservation of the species, with or without the light cues for the changing seasons.[16]

Clocks, clocks, and more clocks. As more animals and plants are studied, the evidence of internal time-keeping mechanisms increases.

Gonyaulax polyedra, a one-celled animal known as plankton (the simplest form of fish food), is still one more example of nature's clocks. *Gonyaulax* creates the luminescence that lovers, on ship or shore, see twinkling below the surface of the ocean. *Gonyaulax* run amok is the "red tide," a poison that is collected and concentrated by clams and other shellfish. And *Gonyaulax* exhibits, like much aquatic life, more than one set of rhythms—it has four.

When its luminescence is temporarily eliminated by giving it an antibiotic, then restored as the drug wears off, the rhythms reappear exactly where they should be, and all four are synchronized with one another once again. The *Gonyaulax* controls its rhythms but is independent of them. It keeps time even when one of its elements is not functioning, and when that one returns, all is reoriented to its original state.[17]

More intriguing still is the response of the plankton to a change in habitat. When the *Gonyaulax* is placed in a laboratory under constant conditions, with the time-giving tides and sun removed, it free-runs. Like de Candolle's mimosa, it redefines the length of its day. Yet its rhythms remain synchronized with one another.

This is quite a system to find in a single cell, but it is not unique. Virtually every cell in a multicellular organism, including many in man, demonstrates such characteristics when isolated in a test tube, or *in vitro.*

The *Gonyaulax polyedra* is a eukaryotic organism. The management of such a cell, and therefore its life, rests in the nucleus. The responsibilities of the nucleus are not only to pass on the genes and inherited characteristics, but also to produce messenger chemicals that run about the cell telling other parts what to do and when. A eukaryotic cell is one with a "good nucleus" (a literal translation from the Greek), as opposed to a simpler, less evolved variety called prokaryotic—*pro* means "for" or "toward"; *karyos* means "nucleus."

Whether animal or plant, fish or fowl, all the life we commonly experience is simply a collection of eukaryotic cells. In structure, eukaryotic cells have more similarities than differences. In each there are mitochondria, which constitute the energy factory for the cell. They take raw materials and produce molecules that release energy when they are broken down. The mitochondria in our cells are identical to those in the cells of a rose bush, an oyster, or an elephant.

The commonness does not stop there. Whales and certain grasses have common enzyme systems in their cells. Again, eukaryotic cells have persistent rhythms and have had them for a long time. Corals from the Devonian period four hundred million years ago show a circadian rhythm at work in forming their skeletons.

While it is now obvious that the rhythms of eukaryotic cells are tied to the physical universe, that their period or length is synchronized to the cycles of the environment, circadian scientists are divided as to where the rhythms and the supporting clock reside, inside or outside the organism.

When researchers report their findings to the lay public, we are inclined to think the scientific process is quite cut-and-dried and that everything can be proved or rebutted. In actuality, scientific tides are often tempestuous. The debate between the partisans for endogenous (rhythms coming from within) and exogenous (rhythms from without) theory has been carried on for several years.

All organisms have electrical pulses. Such medical tests as the electrocardiogram and electroencephalogram measure those pulses and provide information about the workings of the heart and brain. The environment, too, has electrical forces, such as the magnetic fields of the earth. The outside—or exogenous—group argue that these external electrical forces, plus other factors such as the rising and setting of the sun, create a rhythm, and a clock, which the organism follows.

The endogenous, or inside, advocates say that the system is within the cell(s), and that the organism receives time cues from its environment and uses them to synchronize its internal time-keeping mechanism. The environmental forces, they propose, are not the clocks but messengers bringing information that entrains the internal clock.

While the exogenous group can point to the use of the magnetic fields by homing pigeons, which use the fields as a backup navigation system, and to flatworms, whose direction of crawl may be altered by rotating a magnet beneath them, the weight of the evidence comes down on the side of the endogenous group because, very simply, you carry your rhythms within you.[18]

An experiment to demonstrate this: honeybees were trained in Paris to feed between 8:00 and 11:00 P.M., then were flown to an identical bee room in New York, five hours behind Paris time. Completely exposed to all the electromagnetic forces cited by the outside group, the bees brought their own circadian rhythms with them and fed between 3:00 and 6:00 P.M. in New York. If the bees were merely following outside geophysical forces, these forces would have had the bees feeding between 8:00 and 11:00 P.M., in New York as well as in Paris. Instead, the bees brought their own entrained rhythms and fed in New York five hours earlier.[19]

If electromagnetic or geophysical forces composed the clock, they would be at maximum strength at one time of day and at minimum twelve hours later as the earth shielded the organism from the forces or reoriented it. Theoretically, then, at the South Pole, where there is no shielding and the forces are constant, the organism should lose its timing and rhythms. Yet, in constant darkness at the South Pole, the activity of the golden hamster, the sleep movements in bean plants, and the "birth" of fruit flies remain within their circadian patterns.[20]

To verify the conclusion that the clock does not result from external factors, another experiment was conducted. The organisms were placed on a turntable that rotated in the opposite direction from the spinning of the earth. Thus, external forces were obliterated along with any rhythms reliant upon them as pacemakers. The rhythms in the plants and animals persisted. In short, geophysical or electromagnetic forces may act as subtle time cues, but separately or in combination they are not the clock mechanism.

But let us go back inside the cell and explore how the cell beats out its rhythms. At present there are three theories, none of which satisfies everyone. The first, the membrane model, holds that the wall of the cell is the time counter. This fits with our knowledge that alcohol, which affects the cell membrane, is one of the drugs that affects circadian rhythms.[21]

The second explanation posits that the circadian period is the result of interactions between the oscillations or vibrations of the molecules within the cell. The activities go at a furious pace and interact with one another to produce one long rhythm.[22]

The third model, the Chronon, argues that the cell's rhythms are dictated by DNA. Not only does DNA control the offspring of the cell, according to this explanation; it also controls the workings of the cell as a program does a computer. The Chronon theory says that the DNA contains the instructions that dictate the rhythms the same way information is stored in a computer loop; its instructions are "read" sequentially each and every day.[23]

As if three theories were not enough, there are the centrists, who say that the clock is not within one part or another of the cell, but that the cell itself is the clock.[24]

Whatever the individual controls for each cell, when many cells are combined in multicellular organs, they surrender themselves to a coordinated rhythm. In turn, the separate tissues and organs surrender their rhythms to a master clock mechanism. As we rise through the evolutionary scale, the function of the coordinator begins to be displayed in nerve tissue and, in the higher forms of life, in the brain.[25]

Rhythms are, truly, the nature of nature. We have seen how organisms as simple as plankton and as complex as insects, birds, and even mammals share a capability to judge time. As science comes to understand more the underpinnings of nature's intricate balance, in this case the chronometry of many organisms, common traits emerge. And those similarities don't cease as we continue to climb the evolutionary ladder.

They express themselves not only at the cellular level but in behavior as well. At UCLA, a "condominium" for rats was constructed. It contained thirty-six burrows that bordered on a central plaza with treadmills, a dining area, and a bar which delivered either plain water or alcohol flavored with anise.

After a few days in residence, a distinct pattern emerged: at the end of their activity period, the rats would belly up to the bar before feeding time, drink and socialize, and then they would eat dinner. Later, before retiring, they would go back to the bar for a nightcap. On occasion the colony would go on a day-long binge, after which they would rest more, reduce their intake of alcohol, and increase their consumption of water.[26]

Innumerable arbitrary, self-justifying conclusions have been drawn from this tidbit of experimental research. But we will draw none now, save to say that man has more in common with the rest of nature than he is often willing to admit.

With these precedents, let us move on to man. For it is he (or she) in whom we are most interested. And from whom we can learn the most.

We do not, in any real sense, run the world. It runs itself, and we are a part of the running.
—Lewis Thomas

3 Human Chronometry

When a caged mouse clambers onto its treadmill and runs itself into a lather, a softhearted observer might be appalled at the cruelty of it all. In fact, the mouse, far from worrying about the apparent frustrations of its fate, likes nothing better than its daily jaunt on the running wheel. Many mice log over eight miles a night and run on the wheel with singular regularity. With regular exposure to light and darkness, the mouse begins and ends its daily jog at almost exactly the same time each day.

If the mouse is placed in constant dim light, rather than in twelve hours of light and twelve hours of darkness (known in the field of circadian science as L:D 12:12), it continues its former habits for a few days. Gradually its running rhythm changes and a new pattern emerges. The mouse starts and ends its destinationless journey almost exactly thirty minutes earlier each day. If it started at 10:00 P.M. on day one, it eventually starts its exercise again at 10:00 P.M., but it takes forty-nine "mouse" days to lap the forty-eight regular days, since the mouse's "day" is half an hour shorter.

The mouse displays two forms of its circadian rhythms, one entrained, the other free-running. When in L:D 12:12, its "clock" is reset each day by the change from light to dark. The cycle of the light acts as a time cue or time giver and entrains its inherent rhythms to conform to the length of the light-dark cycle, in this case twenty-four hours.[1]

When the mouse lives in constant dim light, minus the time cues of the light changes, it displays its inherent, free-running circadian rhythm of twenty-three-and-one-half-hour days.

Free running is a phenomenon that has surfaced before, first with the mimosa plant, then with the plankton: all organisms have a natural free-running period. The length of this period is between twenty and twenty-eight hours and is entrained by means of a time giver to conform to the solar day. It is known to be inherent, not learned; many organisms display a rhythm at birth, such as baby chicks upon emerging from the shell. If two plants are crossed, the hybrid has a rhythm intermediate between the two ancestors.[2] Provided with a switch to control light and darkness, rats display their inherent rhythm by controlling the illumination of their cage according to their circadian rhythm.[3]

In all organisms except one—man—light is the principal and most powerful time giver: the rising and setting of the sun each day entrains the organism to the temporal environment. Changes in temperature can also entrain certain species under laboratory conditions, and so can the time of feeding and eating; but they entrain the rhythm only if there are no light cues, if the organism is in constant darkness or constant light. Sound can also be used under certain circumstances to entrain certain species.

In man, light is extremely important, but as one would expect from our complexity, other time givers have been found to have effects also. Under certain circumstances we can be entrained by sound, and we have also been entrained on occasion completely to ignore a well-regulated light-dark cycle. It is the present consensus of circadian scientists that light is a strong entrainer but under normal conditions we make considerable use of social time cues such as the alarm clock and the three-martini lunch.[4]

Outside a laboratory, an organism lives in harmony with its environment and is probably entrained by a variety of time givers. The dawn time cue may be followed by the time cue of a good meal on a favorite grain or prey. During the course of the day, a cue may come from the rise in temperature and the fall in humidity. Another meal or the coming of night may also provide cues.

Countless experiments have confirmed the principle of "resonance"; to function best an organism needs to have its internal circadian rhythms in harmony, or "synchronized," with its environment. From fruit flies to mammals, living things bloom better, grow faster, produce more, are healthier, and live longer when the time-giving environmental cycles, such as light and dark, are close to their natural circadian rhythms.

Dr. Jürgen Aschoff, director of the Max Planck Institute for Physiology, is a pioneer in the study of human circadian rhythms. Research on human circadian rhythms is being conducted in a number of locations (including the Albert Einstein College of Medicine in New York and the University of California at Davis), but the largest number of studies have been undertaken in an underground bunker in the foothills outside Munich.

Jürgen Aschoff became interested in circadian rhythms (before the term had been coined) during World War II. He was testing the effect of partial submersion in cold water on human temperature rhythms (there was none). He also worked with birds (finches), and he developed several circadian rules that bear his name as a result of his researches.

After the war he conceived of a completely controlled environment for human experiments and set about constructing it. He knew of tests conducted on humans in caves and was interested in replicating and broadening that experimentation.

Concerned about the effects isolation might have on the subjects, Aschoff spent three months visiting with psychiatrists and psychologists across Europe and in the United States. It was their unanimous opinion that isolation such as he proposed would drive the participants mad.[5] Instead, Aschoff has since discovered, it drives them toward euphoria.

With some help from NASA, Aschoff built his *Tier Bunker*, or "animal chamber." Consisting of two suites, the bunker has walls that prevent the penetration of any vibrations, temperature changes, or electromagnetic forces. Each suite has a sitting room/bedroom fourteen feet square, a kitchen, and a bathroom. The ceiling is an electric "skylight" controlled from the outside. Messages (written only) and food are brought in through an anteroom. The two doors to the anteroom cannot be opened at the same time, although they are never locked, so the participant will not feel imprisoned.

A subject must agree to stay underground for at least a few weeks at a time. He must prepare his own food and is limited to a daily ration of one beer. There is no television, radio, or newspapers, though record players are permitted. Despite these ascetic requirements, there is a long waiting list to get into the bunker and each departing "guest" invariably wants to return to enjoy the pleasant sensation of being on his own natural rhythm.

The only instructions given to those who enter are to lead as normal a life as possible and, unless given permission, to forgo a nap after lunch. A rectal probe reports the subjects' temperatures, and the floors and bed are wired to determine the rest-activity cycle. Other tests are made periodically, and each participant is asked to keep a diary of thoughts, observations, and appraisals of mood and alertness.[6]

Knowing that they will have no means of keeping time, that others will set the length of their day, people have tried to finesse the program. Some have arrived having measured the playing time of their records, but discovered the system failed the first time they fell asleep. Others constantly tested the water pressure from the faucets in the bathroom, thinking it should be greater at night. (The pressure has been adjusted so that it is always constant.) At another similar facility, one subject reassembled a dismantled television set he had smuggled in. But usually, after the first few days the desire to calculate time fades and the game is abandoned.[7]

The subject's first week is spent being entrained to a twenty-four-hour day by either the electric skylight or a gong or both. Then the time cues are removed and a few erratic days follow, as the body tries to establish a rhythm independent of the accustomed time cues.

When this sorting period is over, the subject usually begins to free-run on a twenty-five-hour day, often noting in the log that he or she feels very relaxed and content. After a few weeks of free running, the person is reentrained to a twenty-four-hour day for a week, then leaves the chamber. Nearly two hundred people have now completed stays in the bunker, approximately 30 percent of them women.[8]

A remarkably high percentage of those tested adopt a circadian period of twenty-five hours. While most people realize that they are no longer on a twenty-four-hour schedule, they are generally unconcerned by the change. Rather than being upset or feeling "out of whack," they have a sense of contentment and well-being, a feeling of harmony. Those who have experienced the twenty-five-hour rhythm grope unsuccessfully for a word to describe it, but although they may not know exactly why, they like it.[9]

The most obvious rhythm of man in or out of the bunker is that of alternating rest and activity, the schedule of sleeping and waking usually in a one-third/two-thirds ratio. A second rhythm, and one that works in tandem with the rest-activity cycle, belongs to temperature. As we saw in the Introduction, our body temperature bottoms out approximately two hours before we get up. Let us say that it is 5:00 A.M. Our circadian system has not been inactive (it never is); a few hours earlier, hormones from our pituitary and adrenal glands signaled the rest of our metabolism that the time for activity is approaching, and our body temperature begins to rise.

From about 5:00 to 11:00 A.M., the increase continues, and then it begins to flatten out. At 10:00 P.M. the body's temperature begins to drop and continues to fall during the night until the cycle begins again in the early morning hours: the usual cycle on a twenty-four-hour schedule.

When a person is entrained to a twenty-four-hour day in Aschoff's bunker, the daily low point of body temperature occurs at the end of rest, just as it does in the outside world. During the period of sorting, when the time cues are first removed, the two rhythms of rest-activity and temperature bounce around relative to each other and the person is temporarily desynchronized. After he or she begins to free-run, the rhythms again become synchronized. This time, though, the low point of body temperature is at the beginning of the rest period.[10]

Why the relationship of the temperature and rest-activity

rhythms changes between the twenty-four- and twenty-five-hour day is unknown. Since they are synchronized, seemingly to work together, it is thought to be simply because of the difference between our independent and entrained states. When reentrained to twenty-four hours at the end of the stay, the temperature returns to its usual schedule.

Not everyone remains in synchrony when the time cues are removed. Instead, some people go through the week of entrainment, the sorting period, a varying number of free-running twenty-five-hour days, and then suddenly jump forward to a sixteen-hour day or fall back to a thirty-four-hour day. In either case, the body temperature usually remains on a twenty-five-hour rhythm. These "long"- and "short"-day people continue to free-run with their temperature and rest-activity rhythms out of phase with one another until they are reentrained the week before leaving.[11]

On the outside, such desynchronization can have serious implications, as we shall see, but in the bunker "desynchronized" people free-run with no particular complaint. They enjoy the freedom from time constraints, although they do not enjoy it as much as those who are synchronized. Yet when their rhythms are in the correct relationship to each other, which happens when the shifting temperature and rest-activity cycles run parallel for a few hours, they note in their diaries that they feel unusually well, even mildly euphoric.[12]

People whose circadian period is other than the usual twenty-five hours have unique experiences. Most spend one-third of their time resting and two-thirds in activity, no matter how long their day. Long-day people continue to eat the same amount of food—they are unaware of their elongated activity period—and are puzzled that they are losing weight. In a number of instances, subjects refused to believe it when they were told that their term in the bunker was up. In one case, forty-two days had passed on the outside, but the long-day man inside had lived and counted only thirty. He was quite insistent that a mistake had been made.[13]

Another subject, an artist, had chosen his career so that he would not have to live on a time schedule. Before entering the bunker, he demanded that while he was inside he be permitted to indulge one of his pleasures, a nap. He did it with a vengeance. After the usual preliminaries, he jumped out to a forty-hour day which included a fifteen-hour "nap." It was known to be a nap, and not his nightly sleep, by the regularity of his schedule in the morning: the time from waking until breakfast, from breakfast to lunch, and from waking until nap time changed very little between his twenty-five-hour and forty-hour free run.[14]

How much we like the twenty-five-hour day and how well our systems accept and adjust to it was shown in one experiment: after being entrained by the gong to a twenty-four-hour day, the participant was then entrained to a twenty-five-hour day by ringing the gong every six hours and fifteen minutes. He found the new rhythm so pleasant that it took him nine days on the twenty-five-hour schedule to suspect that he was no longer on a twenty-four-hour day. But when he was put back on a twenty-four-hour day after three weeks on his natural rhythms, he noticed the change immediately. And he didn't like it.

After being left at twenty-four hours for a few days, he was then pushed forward to a twenty-two-hour day. He openly rebelled: after floundering about for a few days, he ignored his time cues and went back to living on a twenty-five-hour day. His other rhythms, however, went other ways and he became desynchronized.[15]

Similarly, if people have experienced the natural rhythm in

the bunker and are then given a small light of their own, they will completely ignore the twenty-four-hour light-dark cycle of the chamber, and will bear the displeasure of resting in bright light and being awake in the dark in order to have the enjoyment of the twenty-five-hour day.[16]

New volunteers are being tested every day, those measured previously are being retested, and the results are continually being confirmed. One interesting—if predictable—constant that has appeared: there is no difference between the schedules of the sexes, aside from a similar incidence of long- and short-day free runners. It would hardly do for half the species to be out of synchrony with the other half; women seem to have, on average, the same free-run period of twenty-five hours.

How these rhythms emerge is unclear. A chick breaking out of its egg manifests an immediate rhythmic balance, as do many animals, but man's circadian rhythms do not appear in one massive burst of synchrony. Rather, they develop over a period of time. While most people think babies sleep virtually around the clock, they are actually awake seven to eight hours a day. They simply do most of their sleeping when we are awake. If allowed to live their own wake-sleep cycle, most babies will latch onto the twenty-five-hour rhythm.

Babies do acquire a rhythm in wetting after two or three weeks. At age sixteen to twenty weeks, the child is entrained to a twenty-four-hour schedule—a sign that this has taken place is that, finally, he will sleep through the night. The body temperature rhythm emerges between six and nine months, but the complete orchestra is not in place until the third year, when the adrenal cycle begins.[17]

At the other end of life's spectrum the circadian rhythms are different, too. A person with a free-running rhythm of 25.8 hours in his twenties might have a free-running rhythm of 24.8

in his sixties. Furthermore, the temperature variations decrease as we age. The body will not be as warm at its peak in the later years, nor will its temperature drop as low. The peaks and depths of our temperature either affect or reflect our performance and sleep; older people, whose temperature rhythms are not as high or low as they once were, have been observed to perform less well and sleep less well than they did when they were younger.

The daily temperature changes are affected by more than the functions of our heart and muscles as we burn up calories and generate heat. The body has two more rhythms that control how much heat we lose and how much we retain.

As with the tongue of the dog and the tail of the squirrel monkey, most of our heat loss is through our extremities, especially the hands and feet. We are more comfortable during the summer in a long dress or long pants with bare feet than in shorts with shoes and socks on; in winter, gloves can conserve a great deal of body heat while shielding only a small percentage of our bodies.

The amount of heat manufactured during activity is far greater than the rise in temperature; the amount manufactured at rest is far lower than the drop in temperature. One of the rhythms that helps keep our temperature within its normal parameters is in our forearms and hands, lower legs and feet. When we are active and manufacturing more heat, the hot blood in our arteries passes to the hands and feet for dissipation; when we are less active, the heat in our arterial blood is transferred to the veins for return to our bodies before reaching the hands and feet.[18]

The second rhythm of temperature maintenance is in the conductivity of the skin. While we are active and our temperature is higher, the skin is more conductive and releases more heat; when inactive and cooler, it preserves the heat. Researchers have also found an annual rhythm in our heat loss and retention capabilities. We apparently have some mysterious and as yet unidentified form of insulation. Experiments have shown that when we are exposed to a temperature of, say, fifty-seven degrees in the summer (when our "insulation" is at its lowest), we will shiver far more than when we are exposed to the same fifty-seven-degree temperature in winter.[19]

Performance rhythms ebb and flow throughout the day, though most of them reach their peaks during the period when the body temperature is high. These performance peaks, however, are not all at their peaks at the same time.[20]

The cycle is as we saw it manifested in George. The sexual hormones flow in the early morning to midmorning; our memory peaks in the morning, but fades as lunch approaches; the mental skills peak in the early afternoon. A temporary drop in many performance levels follows lunch.[21]

We also have a daily rhythm of body rebuilding, when our cells divide to replace the worn with the new. Since people are not partial to giving a piece of themselves for the study of cell division rhythms, there was a problem in acquiring tissue for study in the laboratory. The first and classic study of our rhythm of cell division was conducted on foreskins collected after circumcisions. The peak in cell division was found to be between noon and nine at night.[22]

Studies from the University of Arkansas show that cell division in the skin takes place at night with a peak in activity at midnight. Virtually every tissue in the body has a daily rhythm of cell division, the one exception being the brain, where the cells do not reproduce themselves.[23]

If our skills from mathematics to mechanics follow the rhythm of body temperature, which flows along with the rhythm in our adrenal gland, many of our senses have a rhythm that is the reverse: they are dampened when the adrenal output is high and increase in the afternoon as the adrenal activity decreases.

Again, as we saw in George's day, our senses of taste, smell, and hearing are low in the morning and early afternoon, when the adrenal hormones in our blood are high. In the afternoon, as the hormone levels are dropping, the sharpness of these senses increases. Food tastes better at dinner than at breakfast, which may be why we prefer bland food in the morning, such as eggs, rather than pizza. We are aware in the evening of smells and sounds that went unnoticed earlier in the day. Our sensitivity to pain, however, falls off during the evening; we are most sensitive to the dentist's drill between ten in the morning and six at night.[24]

These complex rhythms are all cued, as with other organisms, by a variety of time givers. As discussed above, light is probably one of the most powerful time givers, although man can be entrained without it.

Natural light penetrates the skin and can reach the brain of certain mammals; however, when men sleep in darkness and are blindfolded during the first three hours of light, some of their rhythms are delayed. It has also been observed that people living in rural areas are more inclined to follow the sun in establishing their daily rhythms than are those living in cities.

Yet we have also seen that a man in isolation with a small light available ignores a light-dark cycle, preferring to free-run on his own. One of the problems in determining the importance of light may be that artificial light isn't very much like natural light. It has neither the intensity nor the same amount of similar wavelengths.

The blind can demonstrate that light has a palpable effect on human rhythms. The blind maintain circadian rhythms, but these rhythms differ from those of the sighted. Their highs and lows are less pronounced than those of people who can see, and the rhythms of the blind tend to drift in relation to each other.[25] In fifty patients blinded by cataracts, there was a reduction in the intensity of the rhythms in their adrenal gland that returned to normal after the cataracts were removed.[26]

Dr. Richard Wurtman of MIT has also reported that blind girls start menstruating earlier than sighted girls.[27] Another study, this one of girls living in the Slovakian mountains, showed that the higher the elevation, and therefore the greater exposure to light, the later the onset of menstruation. Girls living at sea level with less exposure to light than those in higher elevations mature earlier.[28]

Whatever light and other time givers may do for most of us, there are people who simply cannot adjust their rhythms to a twenty-four-hour day. In some cases, people simply resist being entrained to twenty-four hours and lead a free-running life on a longer "day." Some researchers feel that there are many more free runners running about than previously suspected, particularly among those with occupations not requiring a time schedule. In other cases, people become entrained to twenty-four hours, but somehow the internal system processes the time givers so that their rhythms are out of phase with the light-dark cycle.

In his late twenties, Douglas Travers was a talented engineer and physicist who had been fired four times within one year for, literally, sleeping on the job. Ten years ago he would probably have been classified as an outcast, but, luckily, he went to a physician who knew something of circadian rhythms. It was determined that while Travers's system could be entrained to a twenty-four-hour day, the time givers around him somehow skewed his rhythms. He was comfortable and productive only when he slept during the day. The physician's prescription:

sleep in the daytime. Travers now works at night for a high-technology company, successfully plying his trade, having harnessed his unusual circadian rhythms.

Our rhythms are every bit as much our personal property as is our personality or physical appearance, be they as average as George's or as unusual as Douglas Travers's. Everyone incorporates external time cues into his own rhythms, but some people have more difficulty doing so than others. Cases have been observed in which the internal rhythms simply will not shift and remain slightly, permanently, out of phase—Travers is the extreme example, and he had to adjust his world to his rhythms.

The adjustment for most people is simpler and involves an internal process represented by a phase response curve. Everyone has a phase response curve, as individualized as his fingerprints, as the same time cues will evoke a different reaction from each of us. If light commences at six in the morning and the alarm goes off at seven, one person may shift his rhythms in response thirty minutes one way, while another person will shift an hour the other way. As each of us is bombarded by the various cues in our environments, everyone's clock mechanism processes them in a constant direction.

Over a period of time, one's rhythms and external time cues mesh into a constant relationship with one another. Some people shift their rhythms forward, so they are up and about earlier and become day people. Others shift them backward, becoming night folk.

Reared in a society where school starts at 8:00 A.M. and work may start at 9:00, we have always wondered why some are so cheerful, if not more alert, at the opening bell, while it seems to take others forever to warm up; at the end of the day, some have a full head of steam and enjoy socializing into the night, while others fall asleep at the dinner table. Simply, rhythms have given each of them different patterns of activity. What is normal and right for one person is exactly that: normal and right; fighting inherent rhythms is only frustrating. Accept your rhythm as you would your height and eye color.

As we have noted, our natural rhythms shorten with age. Since younger people have longer free-running rhythms, they unconsciously want to stretch their rest-activity cycle and tend to become nightowls. Their parents, whose rhythms have shortened, favor the daytime. Over the weekend, when the time cues of school and work are not in force, many tend to stretch their rhythms by sleeping later and staying up longer. The price is paid on Monday when the rhythms must be compressed back to the regular schedule.

Daily rhythms influence not only normal daily functions but apparently also the arrival and exit from this life. After examining over 200,000 cases of spontaneous labor, 2 million natural births, and more than 400,000 deaths, Dr. Michael Smolensky of the School of Public Health at the University of Texas and others have made the assumption, until further information becomes available, that we even have circadian rhythms of labor, birth, and death.[29]

An unequal distribution of human births over the twenty-four-hour day has been known for some time, not only by the physicians who must deliver the children, but also by observations among mothers. The peak time for the onset of natural labor is at one in the morning; it declines then until about noon. Natural births have their peak around four in the morning, then decline to a low point of 5:00 P.M.[30]

Just as we come into the world in the morning, so are we more likely to die then. The peak in mortality comes around six in the morning, with a secondary peak at about four in the afternoon. Despite the fact that midnight is supposedly the time of goblins and ghosts, the death rate at midnight is the lowest.[31]

Annual birth rhythms are clouded by social customs and religion. In countries with a high Catholic population, for example, where sexual abstinence is encouraged during Lent, there is a peak incidence of births about nine months after Easter.

While normal births do not fit a clearly decipherable annual pattern, birth defects do display an annual rhythm, as reported by Smolensky et al. In the Northern Hemisphere, still and premature births are at their peak in April and May. Certain congenital defects of the heart and the bones also have an annual rhythm. The precise cause is unknown, but there are acknowledged seasonal variations in hormone and other functions important to reproduction that could explain the seasonal variations.[32]

There is evidence that we retain an annual rhythm in our sexual appetite as well.[33] Spring is not a season for lovers after all: suicides increase, ulcers flare, male sexual potency is at its lowest, and this appears to be equally true of women.[34]

Throughout nature there are such strong annual physiological rhythms that control mating seasons and sexual potency that it would be surprising if they were not present in man as well, according to Dr. Richard Wurtman. Although most monkeys can breed all year long, many exhibit a rhythm in the time of birth, and therefore the time of mating. The Asiatic elephant has two peaks of sexual prowess. The changes within an organism can be enormous: the red deer stag shows one thousand times more potency in the fall than in the spring.[35]

And so with man: detailed accounts report the female Eskimo becomes so excited she hemorrhages from the nose and mouth. By early summer, the frenzy culminates in an epidemic of venery when wives and husbands are often exchanged. Neither magistrate nor churchman can put the damper on the Eskimos' enthusiasm—Mother Nature does it by shortening the days.

The Eskimos, however, are the only people so far studied by scientists who do not develop an otherwise strong circadian system. Why this is so is unknown but can be attributed to the absence of light cues because of the long days of summer and long winter nights. (People from other locales maintain their circadian sequences for years in the Arctic, however.)[36]

While the Eskimos' sexual cycles might seem very foreign to the rest of us, there have been reports of similar, although less exaggerated, behavior in the inhabitants of other climates. Diaries of married couples living in temperate zones show that intercourse is more frequent in the late summer months in the Northern Hemisphere. Women first menstruate most often in the autumn.[37] A dim echo of the Eskimos' sexual conduct is felt even in Paris. Studies conducted in that romantic city indicate that male hormones decline from October to April. (April in Paris may not be all that they say.)

Taken together, the evidence indicates that we seem to retain a vestigial sexual rhythm. We may no longer need to, but in another day and age it was necessary to conceive in the fall and give birth in the spring, when food is more abundant and climate more favorable. Birth rates do not clearly reflect the continuation of this tendency, but our sexual behavior does.

(This is not our only vestigial rhythm. There would seem to be no need for the young in developed countries, where food is readily available, to demonstrate rhythm in eating patterns. Yet young children eat more fats during the spring, and during the summer they not only consume more calories but shift their diets away from fats, as they eat more starch and sugar.)

Our sexual appetites vary not only with the season but with shorter rhythms as well. During the menstrual cycle there are marked changes in sexual desire. Desire rises rapidly during the first three days of the cycle and stays high until the fourteenth day, after which it begins dropping. A short secondary period of interest appears just before the onset of menstruation and the beginning of the next cycle.[38] Contrary to popular belief, the menstrual cycle is not 28 days among those in prime childbearing years. In a study of women between the ages of twenty and thirty, Dr. J. D. Palmer found that the average menstrual cycle corresponds to the lunar month of 29.5 days.[39]

Being the most conspicuous rhythm we observe, menstruation is probably a good example of how the harmony of our rhythms works within us. Women have often noted that certain incidents upset their cycle, skewing their rhythms. Similarly, when their rhythm is off, they frequently voice complaints of feeling ill. The same situation pertains to all our other rhythms. By continuing to lead an irregular schedule, for example, we throw our rhythms off and they, in turn, throw us off. We complain that we are not feeling or performing as well as we usually do, and we're probably right—and it's our fault.

We have seen in this chapter many manifestations of circadian rhythms in man, just as we saw in Chapter 2 the ebb and flow of the rhythms in lower forms of animal and plant life. We also examined in the last chapter the current theories about how individual cells beat out their rhythms, and how, as we move up nature's hierarchy, responsibilities for initiating and coordinating rhythms are assumed by the nervous system and, eventually, the brain.

In man the process grows necessarily more complicated. Descartes thought the pineal gland, located deep in the brain, was the seat of the soul; today some believe this gland is the vestigial third eye mentioned by Yogis and mystics.[40] If two sparrows are entrained to different rhythms and we switch their pineal glands, the birds also switch the schedule of their rhythms to that of the new pineal. In man, however, there is no evidence that the pineal functions as a clock, as it does in birds. Rather, it is closely involved in our circadian system through the interaction of various hormones and other molecules in our bodies.

In man (and other mammals) the location of the clock, or master coordinating center, is now thought to be in another area of the brain—the superchiasmatic nucleus (SCN), a group or nucleus of cells located near and connected to the optic tracts.[41]

If the SCN is removed from mice, rats, and hamsters, they lose their rhythm; the normally nocturnal hamster becomes active in the daytime and the female becomes permanently estrous, or "in heat." But removing the SCN does not prove that it is the clock, just as cutting the telephone line does not prove that it is the telephone; its location and attachment to the optical system, however, suggests it has a central if not completely understood role in the circadian system.[42]

To understand our system, we can begin by looking at a cockroach. If we destroy the rhythms of one roach with constant bright light and connect its circulatory system to that of a second roach, the rhythmless roach adopts the rhythm of its companion. The message must be passed from one to the other by hormones. The central clock mechanism or coordinator is probably located in the brain, and the brain harnesses the glands that secrete hormone "messages" for the various tissues and organs.[43]

Man's rhythms are thought to be organized very much like the crew of an eight-oar shell: each crewman represents a rhythm and, for that crew, the coxswain is the master pacemaker entraining the rhythms within the shell. He does so by pounding out the beat. The coxswain's beat is not designed to keep each oarsman stroking simultaneously but to give the crewman in front of him a signal, or time cue. Although all the other oarsmen can hear the coxswain, each receives his cue from the man in front of him. Each oarsman or rhythm is entrained primarily by the one immediately above him; the beat flows from the coxswain to number eight, from number eight to number seven, and so on through the length of the shell.

If the crewmen, or rhythms, are removed from the shell and put in a one-man scull, some will display their own rhythm, just as many organs and tissues do when removed from the body. But when they are in the shell, or body, together, they are organized into a hierarchy of rhythms.

Within us are several crews or subsystems, and the coxswain of each crew, although the leader of his boat, usually pays attention to the coach or the central pacemaker mechanism. Since there are so many rhythms within us, the various coxswains, and even some individual rhythms, listen not only to the coach or the rhythm immediately above them but also to what is going on in the other shells.

In one shell or subsystem, the coxswain is a very unusual person; he speaks a language different from his heritage. By birthright he is nerve tissue, but a good deal of his communication is accomplished, not by electrical nerve impulses traveling along the "wires" of the nervous system, but by hormones secreted into the bloodstream. He is made of nerve tissue but acts like a gland. When he sends out his hormones, they knock on the door of other glands or organs, who in turn go about the business of giving signals to other rhythms throughout the body.

Probably the best known of our glands is the adrenal. We speak of "juicing ourselves up" and "getting the adrenaline flowing," but the adrenal has another function. It secretes a whole slew of hormones calling upon the kidney, the liver, the pancreas, even the walls of our blood vessels, in order to control our blood pressure, the level of sugar in our blood, the amount of water we should discharge or retain, and a score of other conditions. Working with the crews or shells that operate the wake-sleep and temperature cycles, it is the hormones of the adrenal that go throughout the body in the early morning and awaken the various functions to prepare us for the activities of the coming day.

The oarsman above the adrenal is the pituitary gland, and the coxswain over the pituitary is an area of the brain known as the hypothalamus. Some time during the night the master coordinating center, the SCN, sends a message to the hypothalamus, which is brain tissue functioning as a gland. The hypothalamus then sends hormones to the pituitary by a special closed-circuit blood system, and the pituitary responds by arousing the adrenal.

As we proceed through the day, the process slows down: the hypothalamus no longer stimulates the pituitary, and the pituitary stops prodding the adrenal. The level of adrenal hormones in the blood starts dropping in the afternoon and we begin the process of preparing for inactivity in combination with the temperature and wake-sleep rhythms.

This chain, from the hypothalamus through the adrenal and on to the functions it influences, is one of the main axes or subsystems within our circadian system. When it is up and running, our physiology is entirely different from when the system has been shut down for the day. There are other subsystems or chains (such as wake-sleep and body temperature); the chains are coupled not only to the central coordinating area but also to each other. Some parts of the system receive and obey signals or time cues from the outside, while other parts are capable of listening only to signals from within us. Some parts are tightly coupled to other rhythms and some can easily drift about on their own. Each performs in synchrony and delivers the proper physiology, behavior, and harmony at the proper time. It is, clearly, a well-run system.[44]

No one yet knows whether the clock is in the membrane of the cell or the nucleus, or exactly how it works. But why some rhythms are tightly coupled to each other and why others have looser arrangements, how the coxswains and crewmen of one shell communicate with those in other boats, whether there is a master coordinating center in the SCN, and just how all of this is pulled together into the synchrony required of life are questions that, at this time, remain unanswered. We do know that circadian rhythms exist and are crucial to our existence. The persistence and universality of these rhythms make them the third great principle of biology and medicine, according to Dr. Elliot Weitzman of Montefiore Medical Center and the Albert Einstein College of Medicine, a renowned expert on circadian rhythms in endocrinology and sleep. The first basic principle is stimulus and response: for every stimulus we receive, our bodies and minds make an appropriate response. We remove our hand immediately from a hot stove even though we do not consciously perceive the pain until after the hand is removed.

The second principle is homeostasis. Posited by W. B. Cannon of Harvard nearly fifty years ago, the notion of homeostasis describes a state of equilibrium within our bodies. The purpose of the various organs within us—our heart, kidneys, liver, and so forth—is to preserve this constant equilibrium, to return us to it whenever we stray. If we are hot, the body perspires to bring the temperature down; if we are cold, we shiver to manufacture more heat and raise the temperature.

The body is homeostatic in that it does operate within certain bounds. But it is wrong to think of this as a steady state, for as we have seen, we fluctuate over the period of a day. Homeostasis also implies a cause and effect relationship: the reason our body temperature goes up during the day and leaves the "normal" level is not because it is a rhythm, but because we are active physically and mentally.

Many knowledgeable people, consequently, insist the rest-activity cycle causes the rise and fall of body temperature during the course of the day. Yet, as we have seen from those who go into desynchronization in the bunker, these are two separate rhythms, each of which can go its own way. The true source of equilibrium in our body—and the third principle of biology and medicine—is the harmony of the rhythms operating in synchrony throughout the day.

We treat our sleep with indifference,
we treat our dreams with contempt.
—William C. Dement

4 *The Rhythms of Sleep*

George Average, like most of the rest of us, spends a third of his life asleep. Yet if he were questioned about why he does it, or what happens to him while he sleeps, he might think the entire line of questioning mad.

Until Nathaniel Kleitman and his colleagues at the University of Chicago fathered the field of sleep research less than thirty years ago, science had been singularly unconcerned with the phenomenon we call sleep. From the time of the Babylonians, philosophers, poets, and scientists have written off sleep as a blissful escape from the problems of life or, in some instances, as a despicable interlude akin to death. Some religions and cultures believed sleep offered an opportunity for the malevolence in our souls to wander about doing mischief—or worse.

Freud decided that dreams were of value to waking man, but for the most part sleep itself has been regarded as a nuisance to be tolerated. Kleitman and other researchers discovered that sleep had been grossly underestimated, that it harbors two of man's three states of existence, and they subjected it to scientific scrutiny.[1]

Kleitman began his study of sleep by observing people sleeping. He noticed that periodically his subjects' eyeballs darted about furiously under their closed lids. The movements, up and down and side to side, are faster than could be made when awake.[2]

The period of sleep during which these optic gyrations occur is called Rapid Eye Movement (REM) sleep. Kleitman discovered that when his subjects were awakened during the REM episodes and were asked what was happening, over four out of five reported that they had been dreaming. It is now firmly established that it is during REM that we have most of our dreams.

REM is one form of sleep; the other is called, logically enough, non-REM (NREM). Both varieties not only have observable physical manifestations but also internal characteristics that are revealed in the tracing of our brain waves by the electroencephalograph (EEG).[3]

Until 1935, when Loomis et al. discovered that the EEG showed separate and distinct patterns during sleep, sleep was considered a constant, homogeneous state. But it was Kleitman and a graduate student, Eugene Aserinsky, who came upon REM sleep, initiating an important element in the study of sleep: relating the EEG to sleep's various activities. Since the brain generates electrical currents that can be detected, small electrodes are attached to the skin and their input is translated by the EEG machine into tracings on paper. Using these tracings combined with their personal observations, Kleitman and his colleagues were able to map the various stages of sleep.

Sleep researchers have discovered that there is a sequence of events in the average night's sleep that is fairly consistent for most people. Upon first lying down for the night, we drift into a rather drowsy, pleasant state. We are in a relaxed, contented, floating form of consciousness. We have not yet lost touch with our environment, but we pay little attention to it. The EEG traces Alpha waves, which are slower than the waves generated when we are fully alert and active and much like those registered during Yoga and transcendental meditation. Quite suddenly, as if at the flick of a switch, we then fall asleep.

At that moment our awareness of the environment is lost. Our eyes begin to roll slowly in their sockets, moving together or going their separate ways. Our breathing becomes progressively deeper and slower.

The jump from wakefulness to sleep requires only a second, and it is a fleeting moment of transition as the following experiment demonstrates. A subject was bedded down with his eyelids taped open (not as unpleasant as it may sound) and a bright strobe light flashing in front of his eyes every second or two. He was given a small switch to turn off the light. He continually pressed it as the light came on, but suddenly he stopped. The light continued to explode into his wide-open eyes but he did nothing to stop it. One second he was awake and reacting; the next he was "blind" and asleep.

The subject had fallen into Stage I of the non-REM form of sleep (NREM has four stages, REM only one). When in Stage I, a subject who has been under some tension or is particularly tired may experience a convulsive jerk during the first few minutes of sleep. Sometimes the body's flexing is in response to a dream or to an outside stimulus. The movement may awaken the sleeper momentarily, but he returns to slumberland almost immediately.

After a few minutes in Stage I, the descent begins into Stages II, III, and IV. Each of these NREM sleep stages is progressively deeper than the one before in the sense that the mind is further removed from the body's environment and requires an ever stronger stimulus to be awakened. The arrival at Stage II is announced on the EEG by quick bursts of erratic waves known as sleep "spindles," or very high, long waves known as a K complex. This visit to Stage II lasts only a few minutes before the further descent to Stage III. High, long, slow Delta waves are characteristic of this stage.

Compared to the EEG waves of wakefulness, these Delta waves flow easily, even looking sleepy. After about ten minutes in Stage III, still more graceful, free-flowing Delta waves indicate arrival at the bottom, Stage IV.

It is difficult to awaken someone at Stage III, but it is virtually impossible in Stage IV. It requires several minutes to bring children in Stage IV back to awareness. The body is not paralyzed (as it is during REM sleep), but the mind is so quiescent that no commands to move are given. People at Stage IV lie still but not necessarily silently; most snoring occurs during non-REM sleep.

Stage IV sleep is responsible for whatever bad reputation sleep may have, for it is in Stage IV that a number of unpleasant events can take place. Bed-wetting occurs during deep sleep. Nightmares are bad dreams that occur during REM and from which we can be awakened, but night terrors are Stage IV horrors. They occur primarily in children and completely dominate body and mind. A child may act as if he is awake, screaming and thrashing about or even sitting up in bed; and despite all this, he is difficult to awaken. And after finally being quieted and consoled, the child again falls into a deep sleep and has no recollection of the incident the next day.[4]

It is also during deep sleep that sleepwalking occurs. In the movies sleepwalkers are able to execute delicate maneuvers, but in reality the movements of a sleepwalker are exaggerated and awkward. They are in the oblivion of deep sleep, and their clumsiness can endanger them.[5]

Like bed-wetting, sleepwalking is thought to be hereditary. One patient, a chronic somnambulist, told of being at his grandfather's house for a Christmas reunion and awakening one morning in the dining room of the patriarchal manse. And he found several other members of the family had also sleep-walked their way to the room.[6]

Stage IV sleep is also the time when the body promotes growth and maturation. During this stage the pituitary gland secretes Human Growth Hormone (HGH) and the various hormones that stimulate the reproductive system. No matter when an adolescent goes to bed, he will not get his daily dosage of these hormones until he reaches Stage IV sleep, about thirty minutes after retiring. If he stays up all night, he (or she) will get no HGH that day, and it will not be until the next night in which he reaches Stage IV that the hormones will once again be secreted.[7]

After twenty minutes in Stage IV, the body begins to move, signaling an ascent through Stages III and II to Stage I. The second visit to Stage I is very brief and is followed by a sudden jump into REM sleep.

The eyes flicker (their movement can be easily observed by another person) and the facial muscles twitch. The fingers and toes may flutter slightly, but the arms, legs, and body from the neck down are completely paralyzed. The EEG will reveal furious, active brain waves, which explains the need for paralysis. If the body were free to move, it could literally throw itself out of bed. For better or for worse, the dreams begin.[8]

Recent research conducted at Harvard indicates that the entry into the REM stage is triggered by a system in the brain which acts in the same manner as an electric motor with an "on" and "off" switch. Studies in cats have demonstrated that when the "switch" goes on, electrical impulses and chemical messengers travel to the cerebral cortex, the center of higher mental abilities, creating the dreams. When the "on" cells and messengers stop firing, the "off" cells and messengers take over, terminating the dream.[9] Because the control is outside the cerebral cortex, dreams, particularly early in the night, rarely reach a logical conclusion. They have a beginning and a middle, but no end. As if somebody turned off a television, the unfolding plot is left hanging.

The first show of the night is brief. Typically, REM sleep lasts only about ten minutes the first time around. After it is over, the descent through the stages of NREM sleep begins again. The entire trip, going down from Stage I to Stage IV and coming back up, plus the ten-minute short subject in REM, takes an average of about ninety minutes. Depending on the length of a night's sleep and individual sleep cycles, most people have four to seven such trips in the course of a night.

As the night's sleep proceeds through the cycles, the amount of time spent in each stage changes. Deep sleep, Stages III and IV, will progressively diminish and REM sleep will increase, while the other stages remain about the same duration. During the second cycle, adults may not reach Stage IV and have only a modest amount of Stage III sleep. REM increases with each cycle until a grand finale with dreams lasting up to an hour. A few minutes of Stage II follows and, finally, a brief visit at the dreamy, relaxed consciousness of the Alpha waves. Portions of the last dream remain, and perhaps even a momentary, residual feeling of paralysis. (See Figure 1.)

Our dreams are a combination of what we were, what we are, and what we would like to be. Those who are blind from

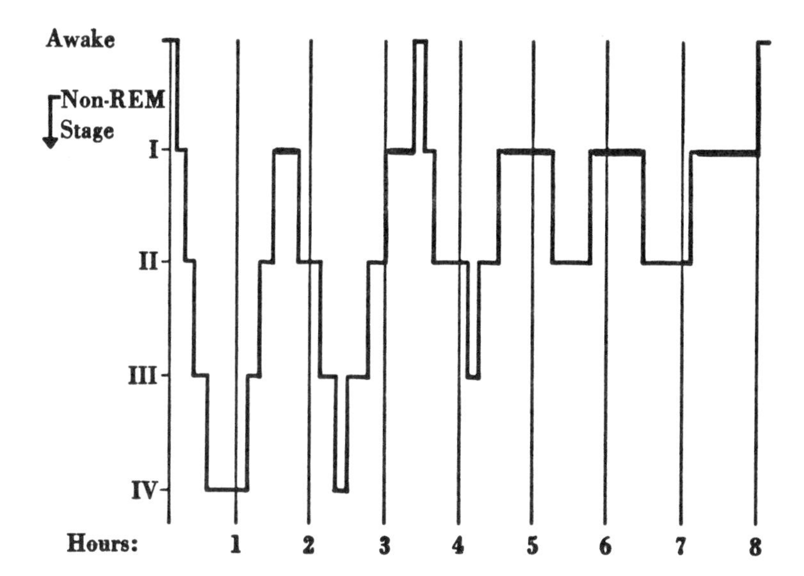

Figure 1. The Nightly Sleep Cycle

This diagram illustrates the nightly sleep cycle of a young adult. REM (Rapid Eye Movement) sleep, the time when we do most of our dreaming, is shown by the black bars.

After a few minutes of wakefulness, the subject falls asleep. He spends a few minutes in Stages I and II before dropping down for a brief stay in Stage III and, finally, during the first cycle, about twenty minutes in Stage IV.

After Stage IV, the subject comes back up through Stages I through IV, then snaps suddenly into a brief, ten-minute period of REM sleep and dreaming. The cycle, from top to bottom and up again, takes an average of about ninety minutes.

During the night, deep sleep (Stages III and IV) decreases and the amount of REM sleep and dreaming increases.

During the fourth hour, the subject awakens for a few minutes but, typically, does not remember doing so the next morning.

birth have no rapid eye movement sleep and, although they dream, they do so only in sound without a visual component. People who lose their sight later in life continue to have REM periods and to dream visually.[10] We may dream about accidents, but we may or may not incorporate a personal injury into our dreams: one subject, a one-time football player who had been paralyzed, still dreamed of playing football years after his paralysis.[11]

The composition of NREM and REM periods changes not only during the course of the night but also with age. Babies spend 50 percent or more of their sleep time in REM. REM sleep progressively decreases until it reaches about 20 to 25 percent of total sleep time in early childhood, a proportion carried through life from that age. As discussed above, children and adolescents have more deep sleep in order that growth hormone be secreted to promote their growth, and the amount of deep sleep each night progressively declines through adulthood. The middle-aged and the elderly may go lower than Stage II only occasionally.[12]

The amount of sleep we require also fluctuates with age. A baby sleeps sixteen to eighteen hours a day, but an eight- to ten-year-old sleeps approximately nine hours a night. From adolescence through the early twenties, the nightly sleep increases by up to two hours. From the mid-twenties through middle age, a shorter sleep time is required, an hour or so less than during childhood. As retirement age approaches, the duration of sleep changes, with people requiring either more or less sleep.[13] A person who slept eight hours at age fifty may, by the time he or she is seventy, sleep only six hours, or stretch the period to ten hours. This is the reason for the frequent complaints in rest homes over a daily regimen of eight hours with "lights out."

As middle age recedes, sleep becomes more fitful. Awakenings are more frequent, are more likely to be remembered, and last longer. The jumps from one stage of sleep to another are faster, and the sleep cycles are bumpier and less continuous than in younger years.

The circadian rhythms are simply aging. Older people do not have the crisp rhythms they once had and do not perform or sleep as well as they once did. This fitful sleep—and the feeling of frustration that frequently accompanies it—often concerns the elderly, but unless there is an abrupt change in sleeping habits, it is probably part of the normal, rhythmic course of events.

Our society often seems obsessed with, but also confused by and ambivalent about, sleep. We relish a good night's sleep and fret if we do not get one. We often become quite anxious about it, since we see sleep as necessary for our competitive survival. It is a means to an end, and we expect it to be efficient or at our command. When it does not obey, we try to flog it into obeying by downing endless nostrums from the pharmacy, grocery, or liquor store.

Like wakefulness, sleep is a natural process within our circadian rhythm. Just as we cannot become fully awake and alert at will under normal circumstances, we cannot sleep on command. It takes time for our body chemistry to prepare the optimum mode for the activity portion of the circadian cycle, and the same is true of sleep. Sleep should be courted, as alertness and performance are during the waking hours of our rhythm, by going about it with a purpose but not an intensity, fostering and respecting it.

Salvador Dali was once asked what he considered to be the proper amount of sleep. He replied that he settled himself in a comfortable chair and grasped a spoon in his fingers. Beneath it he placed a metal pan. As he dozed off, his grip on the spoon would relax and the spoon would clatter into the pan and awaken him. The time it took for the spoon to fall from his hand into the pan was, said Dali, the proper amount of sleep for him.[14] Such is the impracticality of surreal sleep. But many people involved in sleep research entered the field with attitudes almost as impractical.

One researcher calculated that, if a person lived to be seventy-two and reduced his sleep by two hours each night, he would gain an additional six years of "life."[15] With that, or similar notions, many students of sleep began experimenting on the most available people, themselves. They discovered in the long run that their bodies and minds have an inherent sleep-wake rhythm which, after a time, wins out. Under various conditions, it has been shown that people who usually get eight hours' sleep can do quite well on six. But if they are entrained to six hours over a period of time, seemingly long enough to establish a new sleep habit, they immediately return to their accustomed eight hours when the entrainment is removed.

The amount of sleep you should get is how much you want. Your body and mind will tell you and will regulate your wake-sleep cycle. The normal range for adults is between six and ten hours, with a median of seven and a half.[16] Do not be concerned about variations from night to night or at different times of the month or year: if you keep a record you will find that you are remarkably consistent.

If you are, however, getting a great deal of sleep and are still tired during the day, you should consult one of the sleep centers around the country (see Appendix A), or have them refer you to a physician in your area who is knowledgeable about sleep.

If you are a short sleeper but feel and perform well, there may well be no reason to worry. One revealing case is that of a seventy-one-year-old lady being rooster-pecked by her husband because she slept only three hours each night. At his insistence, she went to a physician who specialized in sleep. She was, however, when examined, found to be in excellent spirits and apparent good health, skiing miles across country each day during the winter and hiking a few miles every day during the summer. As long as she could remember, she had slept only three hours a night. She was dismissed from the clinic without further attention.[17]

Some short sleepers, however, are not happy with their patterns. They feel unable to get sufficient rest, perhaps because of an inability to shift their circadian rhythm to conform to the temporal environment. One medical student in his twenties could not get to sleep until four in the morning. He was always in class by eight o'clock, but by the end of the week he felt fatigued on three hours of sleep a night. On Friday and Saturday nights, he would go to sleep at four in the morning and arise at noon, demonstrating a sound wake-to-sleep ratio of sixteen to eight hours.

After initial consultation he was placed in isolation at Montefiore Medical Center. His rhythm was shifted over a period of weeks until he was entrained to go to sleep at eleven and arise at seven. Six months after the entrainment, his rhythms held their position successfully.[18] Similar results have been achieved in other cases.

People simply cannot be entrained to sleep less each night, yet many, like the medical student, can change sleep cycles. The medical student did not have his normal eight hours of sleep altered; his sleep time was simply shifted to conform to the temporal environment. It is relatively easy to shift the phase of circadian rhythm; it is difficult to alter its duration.

A seemingly sleepless night should not be cause for anxiety. It is normal to have such incidents periodically and, contrary to traditional wisdom, a poor night's sleep will not result in poor performance the next day. Irritability may ensue, but experiment after experiment at a variety of institutions has shown that performance is the same as after the accustomed amount of sleep. Accept the loss of sleep philosophically; bear in mind that a slightly longer sleep the next night will make up for the loss.[19]

When a night of sleep is missed, it seems quite natural to take something to promote sleep or to take a nap. Don't. Going back to the original, customary sleep period is best—and without the use of alcohol or drugs.

A nap may actually be detrimental. Studies by the navy show that after forty hours of activity, a two-hour nap results in sluggishness and poor performance.[20] Recuperating from the fatigue of sleep loss is more difficult under such circumstances and the nap should not be taken, especially during the circadian activity mode. Sleep deprivation can be overcome satisfactorily only by sleeping during the sleep phase.

Although one night of inadequate sleep is not harmful, several continuous days and nights of sleep deprivation take their toll. Sensory perceptions are dulled, and reaction time and motor speed are slowed. The ability to memorize is reduced, and irritability increases.[21]

The fatigue of a prolonged restless period follows a circadian rhythm. Soldiers kept on maneuvers for three days feel most fatigued at five in the morning and least tired at six in the evening. Fatigue varies inversely with body temperature—as the body temperature declines, the sense of fatigue rises.[22]

During the first night without sleep, there is a powerful urge to go to sleep between three and six in the morning. To stay awake for a second night is virtually impossible unless the subject is prodded or highly motivated.[23] Interrogators during war have used sleep deprivation, in combination with the stress of the social isolation and general situation, to cause emotional breaks in susceptible subjects and elicit confessions. The crucial factor in overcoming sleep loss is probably physical fitness, as demonstrated by American astronauts, who functioned well despite very irregular sleep schedules.[24]

While deprived of sleep, certain subjects have had psychotic breaks. One man engaged in a "wake-a-thon" for promotional purposes. During the final days of his escapade, he refused to take food or fluid. He was firmly convinced that he was being poisoned by someone trying to prevent him from achieving his goal.[25]

While there are a number of similar documented cases, most authorities believe that sleep deprivation causes psychotic breaks only in people predisposed to mental problems, who would have breakdowns in other stress situations.[26] At the other end of the spectrum from the starving paranoid stands Randy Gardner, a teenager from San Diego, who stayed awake for 264 hours. He suffered no ill effects, and he never even came close to making up the sleep he missed—nor did he try.[27]

One basic rule of thumb for a good night's sleep, as obvious as it may sound, is: do whatever relaxes you. Some people require silence and others need anything from music to a radio talk show to fall asleep. Research has not found that the length or quality of sleep depends on the kind of mattress, firm or soft, or if a waterbed helps. The choice is an individual one; whatever is most comfortable. It is known that hard surfaces, such as a floor, detract from the quality of sleep, as do loud noises. The amount of deep sleep and REM sleep is reduced because of more awakenings, and more time is spent in Stage I.[28]

No ideal temperature for sleeping has been determined, but temperatures above seventy-five reduce REM sleep and cause more awakenings. As temperatures drop into the fifties, the dream content becomes more unpleasant and filled with emotional conflict.[29]

In general, a strange environment interferes with good sleep, and the less disturbance, the better the sleep—thus, we sleep better separately than together. The best thing for promoting good sleep is regular exercise. Athletes have more deep sleep than other people. The most beneficial time to exercise is in the late afternoon, since exercise at night causes stimulation that will disturb sleep, and exercise in the morning does not seem to have the same salubrious effects. People who are gaining weight sleep better than those who are dieting, which is attributable to a survival instinct: the well-fed can afford the quiet slumber, whereas the hungry must be alert for the possibility of securing food.[30]

There are a few rules which, if followed, can help secure a good night's rest. Generally, the bedroom (and bed) should be restricted to sleep and lovemaking; it should not be used as a television room, reading retreat, or dining area. Perhaps even more important, one's attitude toward sleep should be one of serene acceptance. Welcome it as a pleasant part of life, rather than thinking it an unwanted necessity or nuisance. And the establishment of a regular time of awakening also helps develop a regular time for sleep onset by entraining the circadian rhythm.[31] Research from Dartmouth, the University of Edinburgh, and elsewhere indicates that caffeine should be avoided. Those who claim that coffee or tea does not affect them when taken in the afternoon or evening are wrong. They may or may not have trouble going to sleep, but the sleep will be fragmented. There will be less deep sleep, more awakenings during the night, and the total sleep time will be reduced.[32]

On the other hand, a light snack before going to bed is frequently beneficial, and Grandma was right: a cup of warm milk, Ovaltine, or Horlick's improves sleep significantly, not only because of the warmth and calories, but also because all these foster the synthesis of serotonin, a molecule in our brain associated with sleep.[33] (There are different chemical or metabolic pathways for sleep and wakefulness. The brain uses several different neurotransmitters to carry impulses from one nerve cell to the next. Norepinephrine, for example, is associated with wakefulness, serotonin with sleep.)

Drs. Peter J. Hauri and Peter M. Silberfarb of Dartmouth Medical School have reported that two aspirin may also promote sleep if taken occasionally. If used more than three nights in succession, however, aspirin will gradually lose its effect and after two weeks the effect will disappear.[34]

The word *insomnia* is often bandied about in our society but nonetheless is seldom understood. Insomnia is a chronic and continuing inability to sleep which interferes with efficient functioning during the daytime or work period. Between 12 and 15 percent of the population of industrialized countries suffer from insomnia; 20 to 25 percent more have occasional sleep problems.[35]

Many sleep problems are associated with irregular, or out-of-phase, circadian rhythms. During their "sleep time," poor sleepers tend to have higher body temperatures than good sleepers, higher heart rates, more frequent awakenings, and a poor distribution of the stages of sleep. After the poor sleeper awakens, the body temperature often continues to drop. All these symptoms suggest a dysfunction of the rhythms.[36]

The inability to fall asleep at night is not the only form of insomnia. There are also varieties that cause those they affect to wake up too early in the morning (and stay awake) or to awaken for several hours during the night between two segments of sleep.

The best procedure, if you can't get to sleep at night, is to force yourself to get up the next morning and into a normal day's activity; if you awaken too early, stay up a little later at night. In both cases, the rhythms may gradually adjust over a period of weeks.[37] If the insomnia persists, get qualified help from or through a sleep clinic. Sleep is one of the few things that are not improved by trying harder, and if its absence becomes perpetual, a serious situation can develop into a vicious cycle of too little sleep, frustration, fatigue, and a dependence on pills or other drugs.[38]

Just as there are varieties of insomnia, there are many causes, one of them, unfortunately, iatrogenic: that is, brought on by the treatment prescribed by the well-meaning but uninformed physician.[39] Typically, the patient complains of an inability to get to sleep or of poor sleep, or both, perhaps from some new pressure at work or at home, and a hypnotic or sleeping pill is prescribed.

After as few as five days, the prescription loses its effectiveness, so the patient increases the dosage. Then, perhaps the job pressure or home problem is alleviated and the patient decides to forgo the pills. Suddenly he suffers from rebound insomnia, and his body and mind are unable to sleep without the drugs.[40]

Rebound insomnia is frequently much worse than the original insomnia, and the patient consequently increases his dosage once more and may even seek a second type of sleeping pill from another doctor. When the effectiveness of the combination wears off, a third pill may be added. The nightly ritual may also involve generous quantities of alcohol. Patients have arrived in the waiting rooms of sleep clinics gobbling dozens of pills from as many as three prescriptions and washing them down with a fifth of liquor.[41]

Sleeping pills have their place in short-term sleep problems when they are properly controlled and their various actions and side effects understood.[42] Not only do their effects decrease with continued usage, but sleeping pills, like alcohol, do not deliver good sleep. "Pill" sleep is fragmented and jumps from stage to stage rather than following the normal, undulating, ninety-minute rhythm. There may also be frequent middle-of-the-night awakenings.

Withdrawal from chronic sleeping pill or alcohol use is an unpleasant experience. Even though the withdrawal must be gradual, it is not easy; the person does not sleep well for a time, which is upsetting in itself, and he or she may come to believe sleep is impossible without the medication. Furthermore, it is a very slow, tedious process requiring many weeks, often accompanied by physical side effects and unpleasant nightmares.

According to Dr. William C. Dement of Stanford, alcohol and sleeping pills suppress REM sleep. When the alcohol or pills are withdrawn, there is a REM rebound. The REM period lasts longer than normal and the dreams often take the form of disturbing nightmares.[43] It is the apparent necessity to recoup missed REM sleep, when the body is weaned off the scotch bottle, that causes the DTs; the normal barrier that confines dreaming to periods of sleep is broken and the dreams intrude on the conscious state and cause horrible hallucinations.[44]

Science has trouble understanding not only dysfunctional sleep, but normal sleep as well. The National Institute of Mental Health reports that REM sleep may have an integrative function. All that is assimilated or experienced during the day is sorted out and consolidated into memory, according to that theory.[45] Many sleep scientists believe REM sleep has an especial importance to emotional well-being. Other theorists argue that sleep has a restorative function; we all feel better after a night's rest. Recent research from the University of Edinburgh indicates that there may be more active protein synthesis in the brain during sleep.[46] Although this provides a potential clue to the restorative nature of sleep, the accepted fact that some people require ten hours' sleep and others only three prompts the question how one person's restorative chemical reactions could be that much faster than another's. Furthermore, as we have seen, people can be deprived of sleep for a night and, if motivated, perform well the next day.

Sleep does keep us out of trouble, preventing us from being active when we are inefficient. Dr. Loring Chapman of the University of California points out that some researchers feel that our head, like a high-rise building full of elevators, heat ducts, and wires, has reached an optimum size; that there is no more room for additional nerve cells with all the nerves, veins, and arteries that come into the head and go out of it. Therefore, to expand the capacity without increasing the size, nature developed a system whereby the brain runs above its average capacity during activity and below this level while at rest. A limited number of nerve processing cells in our brain function beyond their normal capacity during activity, and they compensate for this by reduced activity during most of the night.

There is not yet a single, definitive answer, but all those who argue the various theories admit there simply has to be some function of sleep. It would not have evolved unless it filled some need—nature wouldn't have made so major a mistake as to force its most evolved being, and every other mammal as well, to some greater or lesser extent, to spend one-third of its life asleep without reason. While other functions have shown extraordinary development in the evolution of mammals, sleep has not changed. It is far more important to everyday life than man has been willing to admit; paying attention to its rhythms and routine can make us happier and healthier.

There is absolutely no question that one can overshoot the stimulation of the endocrine system and that this has physiological consequences that last throughout the whole lifetime of the organs.

—René Dubos

5 Stress

When Nathaniel Kleitman wired a patient to an electroencephalogram and combined this procedure with personal observations of the sleeping subject back in 1953, the turning point in sleep research had been reached. Mankind's giant leaps forward often result from the impetus provided by a similar fusion of two systems, disciplines, or fields of study. The curiosity and dedication of Kleitman and his associates were essential to sleep study, but it was the inspired notion of harnessing the EEG to the "sleeping" brain that led to the evolution of a distinct discipline.

A quarter-century later, we are seeing the promise of a like fusion taking place between the study of stress and the study of circadian rhythms. Stress, rhythms, fatigue, and some diseases appear to be as intertwined as a spider web. Experiments have shown that stress can subvert circadian rhythms; conversely, the disruption of the circadian cycle can cause stress. Fatigue often results in either case, as does disease. The links are often clear.

The relationship between circadian rhythms and stress can be visualized if we see our daily rhythm as a wave with a crest and a trough occurring each day. The horizontal axis, the distance from crest to crest, or trough to trough, is time, the length of the daily rhythm. The height of the wave, from crest to trough, is the amount of stress we accommodate during our daily rhythm cycle. As we travel a normal circadian schedule with its cyclic ups and downs, stress should work in tandem with our rhythms. In other words, the best time to incur stress is when our bodies and minds are prepared for it, when our rhythms are at or near their crest; when we are near the trough, stress should be avoided or minimized.

The regularity of this schedule can be changed by either variable, the time or the stress. If the circadian rhythm is expanded or compressed by a change in schedule, we suffer discomfort and fatigue or worse, since the stress patterns do not adjust instantaneously. If we experience a great increase in stress, it affects our rhythms, particularly if the increase comes at a time in our daily rhythms when the stress is usually low.[1]

One sign of the interdependence of stress and rhythms that suggests their intimate relation is a shared pathway: the hypothalamus-pituitary-adrenal network that composes one of the major groups of circadian rhythm organizers (as we saw in Chapter 3), is also the major venue for stress. Although a common chain of command does not a brother and sister make, it does further suggest an intimate relationship. And there are other parallels. There is a photoperiodic effect on the pineal gland; during the long days the pineal is inhibited. In turn, the inhibition of the pineal leads to active gonads. Stress, too, affects the pineal, by encouraging its activity and thereby decreasing sexual drives.[2] The circadian system and stress reactions share this and other common points, though they are not precisely defined.

Perhaps the most telling aspect of the interrelationship between stress and circadian rhythms is contained within the definition of stress. Stress is the result of any external or internal stimulus that excites the nerve cells of the hypothalamus to release hormones; they stimulate the pituitary, which in turn stimulates the adrenal gland. The hormones are produced at rates greater than would occur *at that time of day* had the stimulus not been present.[3] The two concepts of rhythms and stress could hardly be more clearly linked.

The word *stress* hisses from the mouth when spoken; its sinister sibilance may explain why so many people fear stress. But stress is a necessity. We need the right type of stress in the proper quantity at certain times. We were not evolved for a life of inactivity and too little stress, and underwork can be just as bad as overwork.

Stress becomes problematic when our various rhythms are no longer in synchrony with each other or when too much stress is imposed at the wrong time in our cycle. By the nature of their work, firemen must work irregular hours and have heavy demands placed upon them at irregular times. Consequently, firemen show a high incidence of stress-related diseases such as hypertension, coronary artery disease, and stroke.

Reaction to a wide variety of stress inducers varies with the circadian cycle, and the time at which animals contract an infection can influence the course of the disease. The susceptibility of rats to developing stomach ulcers from electric shocks varies with the circadian cycle: during activity 93 percent developed ulcers whereas during rest only 21 percent became ulcerated. When mice are subjected to white noise, 85 percent of the animals have convulsions at the beginning of activity while none have convulsions from the same dose at the beginning of rest.[4]

There are three types of stress that tax our bodies, minds, and nervous systems. One variety consists of physical stresses such as temperature, either too high or too low; noise, which can irritate us during the day or disturb our sleep at night; constant vibration, often accompanied by continuing noise; and humidity, again either low or high.

There are physiological stresses, such as sleep loss or disturbance, irregular eating patterns, and the adverse effects of alcohol and nicotine.

Finally, we may experience the psychological stress of fear or frustration, or the social or commercial pressures that cause anxiety.[5]

Not only are the detrimental effects of stress compounded when they are imposed on internal desynchronosis; stress can also affect normal rhythms. Stress has been used to produce a disruption of the basic circadian rhythms in the laboratory, and with the resultant disorders in time sequence came abnormal emotional and physiological symptoms. Stress, in short, can cause the internal rhythms to lose their synchrony.[6]

If stress is experienced at seven in the evening, the adrenal may slip out of phase with the other rhythms and perform, for instance, as if it were three in the afternoon. The uncoupling of the adrenal and its normal stimulant may desynchronize the gland for a week. The adrenal output of ulcer victims is typical of this: their adrenal output differs from that of healthy people in magnitude and in its time relationship to the other rhythms of the body.[7]

A person under stress may drift away from the twenty-four-hour cycle of light and social activity. Other rhythms also can be disrupted. Subjects placed in isolation in a comfortable "den" well stocked with coffee, soft drinks, food, and cigarettes do not partake in a random fashion. Instead, the subject eats, drinks and smokes in a rhythmic pattern of ninety-six-minute intervals. This rhythm approximates that of our sleep rhythm and also has been found in certain functions when we are awake. During or after stress, the rhythms of consumption shrink to approximately sixty minutes, which is thought to be a regression to the sixty-minute cycles of infants.[8] When under stress, learning and experience are peeled back, returning the individual to instinctual, unentrained rhythms.

A growing number of physicians in industrialized countries feel that the basic cause of much twentieth-century disease is stress.[9] There can be little doubt that in the last hundred years the environment of Western man has changed far faster than he could be expected to evolve. In addition, the evolutionary process in prehistoric times was accompanied by a natural ventilation of emotions, whereas today a lid is clamped on them.[10] We keep them bottled up inside, while daily we confront the peculiarly modern and difficult frustrations of an unending sequence of deadlines and schedules.

Of all the facets of our lives that create tension and stress, none stands higher on the list than a job. Hard work in itself does not cause the problem. Many people put in long hours and enjoy their work; they not only are engrossed by it but relish it and derive a feeling of satisfaction from it. A study from the University of Michigan shows that doctors, for instance, enjoy good health with few diseases either physical or mental. Automobile workers, on the other hand, suffer from tension, boredom, depression, and irritability. There seems to be a high correlation between boredom on the job and anxiety, depression, and physical illness.[11]

Researchers studying both circadian rhythms and stress point to air traffic control as a problem occupation. Not only do controllers suffer from constantly shifting work schedules, but while on the job they come under intense pressure. Each blip on the radar screen is seen as a person—or many persons—not a plane. The effect of a close call upon the controller is the same as it would be if he were flying the aircraft. Controllers have four times as much high blood pressure and twice the frequency of peptic ulcers as the population as a whole.[12]

The effects of stress vary greatly from individual to individual. The pattern in each of us is a unique characteristic, like our circadian rhythm, but in general the body responds to a stress situation as if it were going into shock, as if it had been seriously wounded and was losing blood. The pituitary secretes ACTH as well as other hormones. Under stress, it stops water loss by inhibiting our kidneys and raises our blood pressure by constricting our blood vessels. All this happens in about two minutes after the onset of stress.

Fifteen to thirty minutes later, the adrenal starts secreting its hormones into the bloodstream. If we sample our blood during the reaction and measure the amount of the various hormones present at each time, an individual profile emerges.[13]

Often the pattern reveals that everyone has a specific target organ. If, for example, we are continually under too much stress and tend to what is called the "angiotensin channel," (angiotensin is the hormone that helps increase our blood pressure), we will be subject to high blood pressure. Other "channels" can lead to heart attacks or ulcers.[14]

C. F. Stroebel conducted a revealing experiment on a group of thirteen monkeys; the results suggest how stress manifests itself in different ways. Stroebel subjected rhesus monkeys to various stress inducers such as heat, strobe lights, and electric shocks. The animals were provided with a lever which, when pressed, would relieve the stress. Eventually, each monkey simply rested its hand on the lever, keeping it depressed at all times regardless of whether the stress was or was not on.

Abruptly the stresses were eliminated. At the same time the lever was also moved so that the monkey could see it but not touch it. All thirteen monkeys became frantic and, under this self-induced stress, twelve of them divided themselves into two subgroups distinguishable by the difference in the rhythms of their brain temperature.

The "psychosomatic" subgroup, Dr. Stroebel discovered, had a brain temperature rhythm that was out of phase with the light-dark cycle and the regular feeding times. Some of the monkeys in this group had a temperature rhythm a bit longer than twenty-four hours and some had a rhythm a bit shorter. Although they continued to perform certain tasks, they were inefficient and developed neurotic and psychosomatic symptoms: some had asthmatic breathing, two developed duodenal ulcers and died, all showed intestinal disturbances and skin lesions. They all drank more than a normal amount of water, and three developed high blood pressure.

The seven members of the "psychotic" subgroup had a brain temperature rhythm that expanded through stages of sixteen and thirty-two hours to a forty-eight-hour period. They were lazy and weak, ceased grooming themselves, lost interest in food, and took naps. Their behavior was bizarre: some spent hours catching imaginary flies, one masturbated almost continuously, three compulsively pulled out their own hair, and all had long periods of despondency as well as disturbed sleep cycles.

When the "security" levers were returned, the two subgroups continued to show different behavior. The "psychosomatics" who had not died of the ulcers immediately resynchronized their rhythms and improved their behavior. (The monkeys in this subgroup who died of the ulcers were those who had a temperature rhythm of less than twenty-four hours.) The "psychotic" monkeys showed no improvement at the end of five weeks.[15]

Like the less evolved rhesus monkeys, humans respond differently to the same stimulus; stress causes different problems in different people. Among the illnesses physicians now associate with stress are high blood pressure, heart attacks, migraine headaches, hay fever and allergies, asthma, problems with the skin, peptic ulcers, constipation, colitis (inflammation of the large intestine), rheumatoid arthritis, menstrual difficulties, flatulence and indigestion, excessive thyroid activity, diabetes, and tuberculosis.[16]

There is some debate as to whether or not mental illness, in addition to these physical manifestations, can be a stress response. Some experts say yes, others no. One physician cited a study of those living beneath the approach to Los Angeles International Airport; people living in the area have 29 percent higher than average admissions to mental hospitals, and those in similar environs near London's Heathrow Airport have a 31 percent higher than average rate.[17] These people had less deep sleep, more wakefulness during the night, and less REM sleep. The question is thus raised whether the cause could be stress, disruption of the circadian sleep rhythm, or both; but, again, there is evidence of an interrelationship between the two.

Whatever the relationship between mental illness and stress, there is little doubt that stress can be linked to heart disease. Historical researches have uncovered no descriptions of heart attack-like deaths until the late nineteenth century; clinical observations begin to surface in almost epidemic proportions after World War I. Middle-aged males were the most common early victims, but the phenomenon began to spread to younger and younger men. The sudden emergence of the illness ruled out a genetic cause; it was decided that heart attacks must be the result of a change in the environment.[18]

Investigations have zeroed in on diet, blood cholesterol levels, lack of exercise, and smoking; and statistics show a correlation between these factors and coronary artery disease

(CAD).[19] Yet, as the evidence was gathered and the case against them mounted, researchers began turning up other studies that made them wonder if there were not other factors as well. The Irish in Boston were found to have a higher incidence of CAD than the Irish in Ireland, who eat more fat than their American cousins.[20] Equally strange, a low rate of CAD was discovered in twelve tribes from a variety of locales around the world who had heavy fat diets.[21] Another inconsistency was observed in Finland—most Finns follow an identical dietary regimen, and yet the residents of some areas of the country have four times the coronary problems as the people in other sections.[22]

A map of coronary disease concentration in the United States in 1950 is composed of mostly white space, but a large number of black dots are jammed together in the northeast part of the country, in California, and in Nevada. A map of the same subject circa 1970 reveals the concentration of dots not only increasing in the Northeast and around the Great Lakes, but appearing for the first time in the Southeast. The same pattern is true of England: as areas have become urban and industrialized, the incidence of heart disease has increased. Inexplicably, the heartland of America, where people eat beef, eggs, and butter, has a twofold to fourfold lower incidence of coronary disease. In two hundred years, Americans have not substantially altered their diet to account for this change.[23]

Today joggers and runners abound, many of them wearing out their track shoes in an attempt to reduce the chances of succumbing to heart disease. Despite their ubiquitous presence, there is no evidence that jogging—or any other exercise—has a direct effect on the heart as a pump, enlarges the coronary arteries of the heart, or builds up other collateral circulation routes in case one of the coronary arteries is blocked and causes a heart attack.

It is known that exercise makes us feel better, sleep better, and that it tones our muscles. That it does have a beneficial, protective effect seems likely. But no one knows just how it does this. As a navy flight surgeon put it, there is reason to believe that exercise will make you live longer and it will make you feel better while you're alive.[24] And there's the infuriating statistical fact that although Finnish males are the most active in the world, they have the highest rate of coronary artery disease.

Whether smoking initiates coronary artery disease is unknown; it is clear that it influences its progress and also its prognosis after a heart attack. Many American physicians have stopped smoking cigarettes, but the rate of heart attacks among them has not decreased. Among British doctors who stopped smoking, the rate actually went up 8 percent.[25] In one major study over a period of eighteen years, among those who smoked a package of cigarettes a day, 17.9 percent had heart attacks; among nonsmokers, the rate was 17.7 out of every hundred.[26]

Although known risk factors have been reduced by an increase in exercise and a decrease in smoking, it is not known whether these changes can account for the drop in mortality during the past fifteen years; this does not mean that there is a decline in incidence. But the ever increasing research into heart attacks has produced some intriguing information.

Ray Rosenman and Meyer Friedman have an active cardiology practice in San Francisco. When they sent the chairs in their waiting room out to be re-covered, they were told by the upholsterer that the fabric was not worn in the usual places. Instead, only the front edges of the seats were frayed. Rosenman and Friedman then determined that the odd pattern of wear was caused by patients shuffling their legs back and forth, sitting forward on the edge of their chairs. This oddity, with a number of other observations, led them to investigate the relationship between behavior and heart attacks. Their researches culminated in their delineation of two personality profiles. Perhaps the best way to understand the two personalities is by a game: a "Type A" is matched against a "Type B" to solve a plausible but impossible puzzle during an allotted time period. After a few minutes of pushing the problem around and sensing its complexity, a Type B personality pushes away from the table and chuckles about the impossibility of the task. His Type A adversary continues to work frantically against the deadline, making attempt after attempt to solve the puzzle. The prize, a very expensive bottle of imported wine, has never been awarded to a winner.

Type A personifies the American stereotype. He is in a constant struggle against the resistance of people, places, and things thrown up before him in twentieth-century urban America. His engine runs at full speed to meet a multitude of deadlines. Type A is not a bad sort of fellow—he has many good traits and is relatively free of neuroses—but the perpetuation of him as the paradigm of success is misleading. As many corporate presidents and successful businessmen are Type B's as A's.

Rosenman and Friedman do not know how or why Type A behavior causes or contributes to coronary heart disease. But membership in the Type A club is a risk factor like smoking, that increases the likelihood of CAD. In some manner it also influences the effects of the other risk factors.[27]

Americans are not alone in their battle with CAD. Japanese who stick by their less stressful traditional life-style suffer less incidence of heart disease than those who adopt the hectic pace of American life. A study of Japanese-Americans shows that those who have become acculturated have an incidence of coronary heart disease two and one-half times that of those reared within their Eastern traditions. When classified by both their upbringing and the degree to which they have abandoned the traditional culture, the most Americanized have five times as much CAD as the least Americanized.

There is a gradient to the incidence of heart disease among Japanese. The low point is in Japan, the midpoint in Hawaii, and the high in the United States. The obvious answer seemed, to the researchers' first glances, to be diet. Then it was discovered that the gradient remained among those with high-fat diets in Japan, Hawaii, and California. Other risk factors such as smoking, cholesterol, blood pressure, and weight were also examined; none, however, could explain the difference. One conclusion seemed that, since Japan has the lowest rate of heart disease of any industrialized nation, the culture of Japan must offer support to cope with the stress of urbanized, industrialized societies.[28]

If the sonority of the word *stress* is slightly misleading, its usage is similarly so. People tend to dwell on its destructiveness rather than its usefulness, probably because we are oblivious to the ways it can help us; we grow conscious of stress only when it impedes our functioning. Stress's dual role in our lives renders it not unlike an impenetrable black cloud, the composition of which is not quite discernible, but which, because of its color, we think is something to avoid.

To balance this, to give us something to grasp and use to our advantage, Dr. Thomas H. Holmes of the University of Washington speaks not of stress but of Life Change Units. Instead of wondering at the awesome black cloud, let us inspect its parts, as Dr. Holmes has arithmetically displayed them. (See Appendix B for a complete list.)

Each change in our life, from the death of a spouse (100 units) to a minor violation of the law such as a traffic ticket (11 units), receives a certain number of Life Change Units, or

LCUs. The total accumulated during a twelve-month period suggests the probability of an individual's experiencing an illness or accident. If the score is over 300, the chance of having an illness or accident is 80 percent and it is likely to be a major one.

A serious illness or injury is in itself a change in our lives (53 units). Disease, according to Dr. Holmes, is not scattered at random throughout the population: 30 percent of the people have 70 percent of the disease, since those afflicted tend to have multiple illnesses in a variety of organs.

These patterns of illness continue throughout life. People who are often sick are likely to have had sickly parents; the illnesses begin early and go on. School-age children have the LCUs of their parents until they strike out on their own. During the teen years, independence begins in LCU accumulation (as well as on other fronts). Risk-taking behavior in teenagers does not predict accidents as well as the LCUs. LCU calculations are three times as accurate as other predictions.

Many events with low LCU value are characteristic of younger people, while the elderly have a lower number of events with high LCU value. Divorced people, in the year following the event, have twelve times greater probability of illness, than the general population. The two best predictors of illness among the middle class are (a) residential change and (b) a first illness followed by a second.

While people with low LCUs do get sick, they do so less often and less seriously. Implicit dangers seem to await those who have unstable life patterns with many changes, from simply taking a mortgage to having one foreclosed, from a business readjustment to a change in financial status. The change does not have to be negative: the effect of fighting less with one's spouse, for example, logs the same number of LCUs as fighting more; a change for the better in one's financial status equals a change for the worse; a demotion at work is the same as a promotion. If and when the 300-point figure is surpassed in one year, the odds that cancer, heart attack, or schizophrenia will strike are great.[29]

As human beings, it is our nature to grasp tangibles more easily than a concept such as stress. Changes in our diet, smoking, cholesterol intake, and exercise seem like preventatives and we cling to them as guaranteed cures. Much more crucial to our well-being, however, are such uncontrollables as conflict with our partners, professional and marital; unasked-for life changes (death of a close family member, retirement); or changes that we seek, such as marriage or divorce.

LCUs provide a look at what has happened in the past, what is happening now, and what we may expect should we elect to pile up more points in the future. A life change is a stimulus that causes a reaction, and too much stimulus and too much reaction may cause a disease or accident as we add points to our yearly ledger.

After past or anticipated changes in units, we can help to prevent illness by following certain steps. According to Dr. Holmes, we should do the following:

- become familiar with life events and what they require;
- put the list up and look at it several times a day;
- recognize when a life event happens;
- try to identify the meaning and experience;
- think how to adjust to it;
- make no impulsive actions or decisions;
- anticipate and plan whenever possible;
- pace yourself;
- look at the accomplishment of a task as part of daily life and keep it up, day after day;
- remember, the more life change, the more illness (300

LCUs or more, an 80 percent chance; 150–199, a 50 percent chance; less than 150, a 30 percent chance).

Investigations show that certain attitudes are associated with certain diseases, and those who have the same disease or illness espouse a common perception. A case in point is Reynaud's disease. It is a problem of the blood vessels in the extremities, of peripheral circulation. The blood vessels can become so constricted that oxygen does not reach the tissues; gangrene and amputation may follow. People affected will characteristically describe themselves as threatened, as having to be on their guard.[30]

Just as certain attitudes are associated with certain diseases, so are certain attitudes with recovery. Religious conviction or faith, membership in a majority rather than a minority, and the presence of genuinely supportive family members cause factors that give a patient a better chance to recover. As simple as it sounds, recovery is also enhanced by the belief that life is satisfying and productive or that one has something to contribute.

Nature's laws are immutable, especially those we do not fully understand. We see man's prescriptive law and regulations all around us, and we fall into the trap of thinking we can order nature. But despite judges, boards of directors, and traffic cops, water boils at 212 degrees and freezes at 32. The miracle of our body is that we can tolerate a certain amount of abuse and change, that we adapt to life below sea level or at twelve thousand feet. But we must use stress and our rhythms to our advantage, accepting stress and work as the proper activity at certain points within our daily rhythm. And we should realize that there are certain immutable natural laws that we violate at our peril.

6

At Work and at Play

We have seen how our circadian rhythms constitute a framework within which we live. We exert ourselves when awake, and stress helps us to deal with the exigencies of the day. We retire for the night, when sleep recharges us and prepares our bodies for another day. Such rhythmic patterns are being identified in laboratories around the world.

We have seen a few exceptional cases so far, cases that have played cameo roles in large measure only to demonstrate how the exceptional few have difficulty adjusting to the normal, and how the normal can have trouble with the unusual. Now let us look at commonplace, everyday activities and learn from circadian science what we can about how we work and play.

The professional gambler is not, of course, an everyday character, but his life-style is revealing. He wears a visor, says the stereotype, and rolls up his sleeves; if the real card shark is to live up to his reputation and ante up all night long, he would be well advised to adhere to a schedule of sleeping all day, so that he will be at his best when it comes time to cut the deck and his opposition—the nine-to-fivers he plans to take advantage of—will be fading rapidly. Our ability to multiply quickly and accurately varies with our body temperature. So does our manual dexterity, our ability to sort and deal cards.[1]

The racetrack bettor, too, would be wise to bone up on circadian rhythms. As the sun rises over the track, many a railbird can be seen watching the morning workouts, stopwatch in hand. Perhaps such industriousness ought to be rewarded during the afternoon's betting but don't count on it. Morning glory horses can run like the wind at dawn's early light but can also become sluefooted when called to the post in the afternoon. Jet lag, too, affects them. The prudent trainer and bettor would do well to wait for their horse to become well adjusted to his new schedule.

If man's and other animals' efficiency varies throughout a normal schedule, then the havoc wrought by an ever changing schedule causes even greater variations in productivity.

One current badge of honor in certain parts of the business community is to be termed a "mid-continent man." He is a person who closes a deal in Houston on Monday, sees a client in Boston on Tuesday, then jets to Los Angeles on Wednesday for a conference. His home is a plane over the middle of the continent.

A similar breed are the mid-Atlantic men who are constantly crisscrossing the ocean. An official of the World Bank committed suicide in his London hotel by swallowing sleeping pills and aspirin; it was later revealed that he had made eighteen transmeridian trips by air in the previous twenty-three days. His wife testified that when he arrived home, he could hardly stand up because of lack of rest. While other factors are incalculable, the toll taken by his never ceasing travel can only have hurt.[2]

What is true of the businessman may be even more true of the Kissinger-esque diplomats who practice shuttle diplomacy. Physicians attending them have reported sleep problems, ulcers, and disturbances of the intestinal tract.[3] While stress seems to deserve the blame, we have seen how stress and circadian desynchronosis are intimately related. To the businessman and the foreign serviceman add the military; in recognition of the problems caused by the quick change in time, the army is now conducting a major study about overcoming jet lag in troops.[4]

One does not need to fly across time zones to encounter circadian problems. A visit to the pressroom of almost any major newspaper will provide a case in point. Since most large metropolitan newspapers hit the newsstands in the morning, they must be printed at night. Given the confused rhythms of the pressmen who work all night, it is no wonder that the production manager of one paper has described the breed as eternally drowsy, accident-prone, unable to remember something from one moment to the next, irritable, and constantly complaining of upset stomachs. Had they just disembarked from the Red-Eye Express from Tokyo, we would think they had jet lag.

The newspaper and printing industry is the largest private employer in the United States, but there are many other industries that for a variety of reasons ask their employees to work shifts. Policemen, firefighters, hospital workers, and others, in the service sector must make their expertise available around the clock. The paper, steel, and petroleum industries rely on technology that demands that the machines be run continuously. The continuing energy crisis may increase this trend, as companies elect to work nights in order to take advantage of lower utility rates. At the present time, more than sixteen million people (20 percent of the labor force) are employed in shift work.[5]

The basic problem with shift work—and other pursuits that require schedule shifting—is that the rhythms of body temperature and the adrenal hormones are not synchronized with the schedule. This balance is, of course, crucial to performance and alertness; both factors parallel body temperature and adrenal hormone levels respectively.

The cause of the desynchronization in shift workers is the disruption of the wake-sleep rhythm. Poor sleep is the most common complaint among shift workers. A worker on a day shift who gets 7.5 hours of sleep might go on rotation and get only an average of 6.5 hours, and even less when he works the midnight-to-8:00-A.M. shift.[6]

A study of nurses suggests that the disruption of sleep in the real world is far greater than the disruption people confront in controlled laboratory conditions. The nurses who slept during the day found their sleep cycle highly fragmented, resembling a series of naps. Sleep deprivation ran as high as 50 to 60 percent of normal. A second study showed that rotating shift nurses were significantly more confused, depressed, and anxious than other nurses.[7]

When one is placed on shift work, the body temperature rarely adjusts. Although after a period of time the rhythm gradually shifts to its usual position parallel to the wake-sleep cycle, the temperature does not drop as far during sleep nor does it rise to its customary height during wakefulness.[8]

One reason for this is that shift workers find it extremely difficult to ignore social time cues around them. They often depart from their schedule on their days off; if they are working the night shift and normally sleep in the day, they may forgo their adopted sleep schedule to rejoin their "normal" family life with their spouses and children.

The literature on circadian rhythms contains few cases of shift workers who have completely adjusted their temperature rhythms. One successful adjustment belongs to a pair of German bakers who worked only at night. But they also preferred to be active the entire night on their days off and consequently maintained a constant sleep-wake rhythm.[9]

Another problem in shifting body temperature appears to be in the mechanism of the rhythm. When an individual accustomed to a normal schedule is placed in an isolated environment and his rhythms are shifted by twelve hours, two to three weeks are required for his body temperature to adjust to the abnormal shifted schedule. But when he is returned to his "normal" routine, his body temperature realigns itself in only two to three days.[10] Although such a person's temperature was successfully shifted in the artificial environment, the underlying mechanism in the central nervous system never forgot where the rhythm was supposed to be.

The problems of shift work arise less from working at odd times than from the constant rotation of work schedules. Out of some ill-informed sense of fairness, employees are shuttled from the day to the evening to the night shifts in consecutive weeks. The result, of course, is an utterly confused body clock.

The rotation should occur at intervals of no less than a month, preferably less often.[11] That way the body can have time to adjust to its new routine; the temperature rhythms can resynchronize with the wake-sleep rhythm dictated by the employer. Workers who are permanently assigned to an evening or night shift complain far less than those who make weekly stops at the different stations.[12] Rotation, then, may be the real nemesis.

The problem is not simply a cross to be borne by the individual but a problem for business as well. Shift work imposes economic losses: most accidents occur between 11:00 P.M. and 2:00 A.M.; and studies have shown that errors in accuracy (by meter readers in one case) occur most frequently during the early morning hours, peaking at 3:00 A.M.[13] Almost invariably, workers on rotation not only have the poorest accident record, but their productivity is also the lowest. Finally, a wide variety of personal problems often arise from the shifting schedules, adding costs that are difficult to quantify but are by no means negligible.[14]

(Some people adapt to shift work better than others. Extroverts, for one; others who are less neurotic but more impulsive seem to adapt better. Those who use fewer medications also make the adjustment more easily, as do those who use alcohol to induce sleep.)[15]

A positive attitude toward the job—and life—helps people shift their work hours. A cooperative spouse can make an enormous difference. The predisposition of a potential employee is very important, if hard to gauge. At one oil refinery the workers on the night shift actually had a lower accident rate and fewer illnesses because they liked the hours, had small congenial groups and more job identification and satisfaction.[16]

Repeated shifting can be very debilitating. Organizations using shifts should examine the human and accounting costs of health problems, performance, output, accidents, low morale, and absenteeism. Some illnesses can be exacerbated by shift work, especially if the rotations are frequent. Insulin and its relation to blood sugar has a circadian rhythm, and it is important that diabetics remain on a regular schedule. Epileptic attacks follow a rhythmic pattern, with some patients having nocturnal seizures with a peak between ten at night and midnight, while others peak at six to seven in the morning. Shifting schedules affect them—and the efficacy of their medication. People with hypertension and cardiac problems should also avoid rotational shift work. [17]

For workers unencumbered with such problems, there are

other physical ailments that may occur. Ulcers have been found two to eight times as frequently in shift workers as in those consistently on the day shift. Even those who work odd hours on a fixed shift are absent more often than day workers and frequently complain of respiratory or digestive disorders.

There are some steps that can be taken to improve the situation. The following recommendations for the employee come from Charles M. Winget of NASA:

- perform the most important tasks as soon as possible after waking;
- avoid working with power tools or driving a vehicle when feeling disoriented or sleepy;
- eat the largest meal of the day as soon as possible after awakening—this decreases the caloric intake slightly and avoids overloading the system with food;
- decrease intake of salt, coffee, beer, wine, and hard liquor;
- use only prescribed medication and inform the plant doctor.

For the employer, it is advisable to have workers on a fixed shift if possible; if not, keep them on one shift for at least three weeks and preferably four. Second, employees who might have problems with shift work should be screened out by doing the following:

- eliminate those candidates who are likely to have health or emotional problems (accomplish with interviews and tests);
- screen out any employees required to take medication, such as insulin, whose rhythms can be affected by reversal of the day-night cycle (accomplish through interviews using knowledge of drugs that are time-oriented);
- determine if an employee has trouble sleeping under normal conditions (tests can be made with small EEG monitor-

ing devices recently developed)—anyone who has a radical sleep problem should not be considered;

• assess the performance and alertness outlook of prospective shift workers (temperature-recording curves and urine excretory surveys can be used);

• prepare the employees for the problems they may face and help them cope with these problems by reviewing the list of items the employee can follow to reduce the strain of shift work.[18]

In the past we have paid little heed to the problems of shift work. There are indications, however, that we are about to do so. Recently a sixty-two-year-old cheese maker in California suffered a heart attack that forced him to stop working. In a hearing before the California Workmen's Compensation Appeals Board, his lawyer argued that disruption of his daily rhythms from rotating shift work had created excessive stress that may have contributed to his heart condition. Expert testimony was given by Dr. Donald Tasto, who headed a study of the problems of rotational shift workers conducted at Stanford Research Institute International.

Those who work day schedules have adjustment problems, too. The normal daily routine of eight to four or nine to five favors those who reach high gear early in the day. Others may seem slow or lazy to their employers because their hours of maximum efficiency don't coincide with some of their energetic coworkers'.

One solution may be *flexitime*. Developed in Germany in 1967, flexitime spread across Europe in the following years and crossed the Atlantic to the United States in 1972. Now nearly 13 percent of all nongovernment organizations with fifty or more employees use the system.[19]

Flexitime permits the employee, within certain specified

limits (they vary from employer to employer), to select his or her working hours. If the usual work schedule is eight to five with an hour for lunch, the employee may come in, for example, between seven and nine and depart between four and six, so long as he works the required eight hours.

Flexitime is not suited to all organizations, but those that have adopted it report a substantial increase in job satisfaction. The resultant increase in employee morale has led in many cases to lower turnover, less absenteeism and tardiness, and an improvement in recruiting new employees. Other firms report flexitime increases productivity, probably because the employee selects a time when he or she is functioning best and less time is spent settling into the work routine—the workers arrive at different times and are less prone to discuss last night's TV shows. The employee is more likely to complete a task before leaving, rather than setting it aside until the next day. Flexitime also gives added coverage at no added personnel costs to those companies who telephone other time zones.

Flexitime, however, may require added supervision and cause difficulties in scheduling. To date it has not been possible to determine the "cost-benefit tradeoff" of flexitime, but thus far the evaluations have invariably reported that the benefits of flexitime endure.[20]

If flexitime is not yet a constant in the typical American workplace, there certainly is at least one: coffee. It is the great American upper; certainly, at least, the most commonly used legal one. It is indeed a stimulant, as it increases the heart rate, raises the blood pressure, calls upon the liver to put more sugar into the bloodstream, and stimulates the central nervous system; hence the problems coffee causes for sleepers.

Some early research suggests that coffee may make us more prone to cancer; if too much is consumed in the morning, it can disrupt the circadian time-keeping mechanism.[21] However, the main reason to avoid it is not the specter of cancer, but rather the fact that we don't need it. Our hormone system is doing the same job coffee does. In addition, the body, when at work, struggles to retain water, and coffee is counterproductive: for every cup of coffee drunk, a cup and a half of water will be passed, thereby dehydrating the body.

Coffee's reputation as an agent of sobriety is overrated. It stimulates the mind into thinking it is alert, but it actually does not reverse the other effects of alcohol, such as reduced reaction time. It might be better than nothing, but not much better, and it certainly shouldn't be relied upon.

On the other end of the scale is our biggest downer, alcohol. In spite of the pleasant sensation a drink or two seems to deliver, alcohol is actually the same kind of central nervous system depressant that barbiturates, general anesthesia, and minor tranquilizers are. The restraints on speech and behavior disappear; the sense of fatigue decreases; visual acuity is lost and reaction time slows; memory, concentration, and insight are impaired.[22]

The notion that heavy consumption of alcohol is a modern phenomenon to be attributed to advertising or an industrial society is unfounded.[23] Per capita alcohol consumption in the United States went up 30 percent between 1954 and 1970, but even today America ranks only seventh in the world and on a par with its 1900 consumption level. Number-one France consumes two and one half times as much, while Ireland, the legendary home of the toper and the red nose, is sixteenth. The two areas of greatest per capita consumption in the United States are the District of Columbia and Nevada; Alabama ranks last.

Alcohol, like many other drugs, has a circadian rhythm of effectiveness. The "hair of the dog" first thing in the morning can have quite a bite: a drink in the morning has several times the effect of a drink taken in the afternoon. The body is simply not geared to breaking down the alcohol molecule in the morning, so more alcohol circulates in the blood.[24] The human metabolism rids itself of alcohol fastest between two in the afternoon and midnight, and the traditional "one for the road" can be a very dangerous drink.[25]

One of the great secular triads of Western man is that of wine, women, and song. The effects of the song are, as yet, undetermined, but the combination of wine and women is overrated. Prolonged heavy consumption of alcohol diminishes male sexual prowess. Inhibitions may be lessened, but in the long term the libido is also reduced and the function of the testicles eroded.[26] The evidence substantiates Shakespeare's observation that alcohol "provokes the desire, but takes away the performance."

Alcohol prolongs the daily phase of the circadian rhythms so that "the morning after" arrives with an aura of desynchronization. In addition, alcohol disrupts the composition of sleep rhythms by depriving the user of the proper amount of REM. Being deprived of REM sleep from alcohol, or for any other reason, reduces concentration, impairs the memory, and brings tiredness, anxiety, and irritability.[27] Unfortunately, the only proven remedy (and even then only for a few of the symptoms) is another drink. It is a poor solution that provides only temporary relief and actually compounds the problem.

The impulse to take a drink in the evening is not solely the result of sociological programming. After a day of work, the body craves fluid and some kind of oral satisfaction. During the day, blood pools in the legs, and the system needs a bit of a stimulant.[28]

Alcohol seems to provide all the answers on a temporary basis. But you can avoid it; to reduce the desire for evening cocktails, a glass of water, which solves the dehydration problem, can be "chased" with a juice high in sugar (orange, apple, or grape) for a quick energy lift. Scientists say that a few minutes spent prone is also invigorating. Maybe even suggestive.

In America one category of books perennially outsells all others: cookbooks. Ironically, diet books also rank near the top. Perhaps one reason for the popularity of food and avoidance-of-food books is a trick that nature has played on us. As we have seen, our senses are more acute in the evening, and consequently food tastes and smells better. Yet rodents who consume a certain number of calories at the beginning of activity lose weight while those who eat the same amount at the end of activity put on weight. The same is true of man.[29] The logic implicit in this observation: we should eat our heaviest meal early in the day and settle for lighter fare in the evening. (The extreme of this theory is contained in Buddha's advice that his disciples should eat nothing after noon.)

The time of consumption appears to be as important as what we eat, because of the manner in which food is metabolized.

Recently, it has been discovered that the electrical charges pass from one nerve cell to another by means of a neurotransmitter (such as serotonin, which is associated with sleep). These neurotransmitters are little molecular bullets fired by one nerve cell into the next.

Sleep nerve bullets are different from those present in waking hours. Experimental animals, shot with a dose of sleep neurotransmitters when awake, go to sleep. This switching of transmitters, or bullets, is part of the overall change in body chemistry between activity and repose.

Breakfast, according to old wives and scientists, is the most important meal of the day. But the dietary regimen practiced by most of us is exactly backwards. Instead of cereal or starch, breakfast should be high in protein. The chemical pathways of activity are facilitated by protein. Logically enough, the evening meal should be high in starch and low in protein to help the changeover to the sleep transmitters.

The time when fuel is taken on is one of the body's rhythmic factors; the time of the fuel's expenditure is as well. Regular exercise acts as a time giver for the circadian system. It is advisable to exercise at approximately the same time each day, preferably in the afternoon. The more strenuous the exercise, the later in the afternoon the better. That way the peaks in the heart, lung, and circulatory functions (between 3:00 and 7:00 P.M.) coincide with the demanding activity.[30]

Not only can we maintain synchrony by using our knowledge of circadian rhythms, but performance can be maximized as well. The advice given George Average in the Introduction applies to the rest of us as well. The memory is best in the morning, so taking examinations or giving testimony is best done then. Mental skills peak at mid-day, so then is the optimum time to ratiocinate. Solving mechanical problems is easiest in the afternoon.

The memory obviously does not take a vacation after noon, nor does mechanical aptitude fall off so sharply that no one can drive a car at night. But given the opportunity to control our schedules, we will not only do better at times of peak efficiency for a given function, but we will probably enjoy the daily chores more. Knowing our cycles can help us work best and play better.

7

In Sickness as in Health

Healthy people can benefit from a familiarity with circadian science; the sick can as well. The body's internal cycles can be both a cause of and a cure for disease. Arhythmic behavior can be a symptom of certain illnesses, and its observation has led, in several cases, to the discovery of previously unexplained medical problems. But before discussing individual diseases, let us look at what is perhaps the most obvious and most important link between medicine and circadian science.

As discussed in Chapter 6, the body has different levels of susceptibility to the drug alcohol at different times of day. Other drugs, too, have varying effects depending upon when they are ingested. A medication that affects the adrenal cycle, for example, varies in efficiency with respect to the body's cycle of hormone secretion. Allergic reactions to outside stimuli are greater at certain times of day than at others, and the medicaments prescribed for them have characteristic cycles. Circadian science has revealed that time of day is a major factor in the potency of most drugs and should be considered in the evaluation and usage of a medication.

When pharmacists attempt to determine the safety of a drug, the FDA requires that they establish its LD_{50} level, or the lethal dose necessary to kill half the animals tested. This is not as illogical as it might sound, not least because a number of drugs frequently prescribed at the corner pharmacy originated as poisons. Digitalis was introduced to Western medicine by primitive peoples who poisoned the tips of their darts or spears with it; today it is used in the treatment of heart disease.

A factor that is not yet considered in the LD_{50} calculations is the time of day at which the drug is ingested. Experiments demonstrate that a dose of a drug that kills a high percentage of test mice at one time of day may kill none of them at another.[1] Digitalis has several times the effect in the early morning that it does when administered at other times.

Cortisone approximates the effects of the body's adrenal hormones. It is often prescribed as a treatment for arthritis, and its purpose is to amplify the effects of the adrenal hormones. Ideally it should be administered in the morning and only every other day, but generally the patient is given no such strictures. In fact, he is often instructed to ingest the drug at intervals throughout the day.

Cortisone and other drugs that affect the adrenal system can have side effects, particularly if taken for a prolonged period. Weight gain can result, and young people's growth can be stunted. Such side effects can be reduced by taking the drugs only in the morning; this practice, though, is not yet widespread.[2]

The rhythms of most allergies are exactly the opposite of those of the adrenal cycle. The average person's sensitivity to house dust, pollen, mold, feathers, and grass peaks at eleven at night and bottoms out between seven and eleven in the morning. People who are subject to allergic reactions and have gardens would do well to do their weeding in the morning. Likewise, most asthma attacks, which are thought to be allergic responses, occur in the early hours of sleep.[3]

The effectiveness of drugs taken to counteract allergies matches the rhythms of the sensitivity: an antihistamine, such as cyproheptadine, lasts only six to eight hours if taken at night; taken at seven in the morning, it lasts fifteen to seventeen hours.[4]

Aspirin is probably our most commonly used drug. It, too, shows a retention rhythm. A dose ingested at seven in the morning lasts twenty-two hours; the same dose taken at 7:00 P.M. is excreted more quickly and lasts only seventeen hours.[5]

A surgeon's worst enemy is the patient who, though anesthetized, does not lie still. Formerly, curare was used to relax a patient's muscles and prevent him from moving about on the operating table, but sometimes it worked too well: patients were lost through paralysis of the respiratory muscles. The explanation lies in the timing of the drug's introduction into the patient's system. In rats there is a double daily rhythm of susceptibility to curare; identical doses gives a reading of LD_{50} at either midnight or noon, but 60 percent of the rats die from a dose at eight in the morning or evening.[6] Another anesthetic, halothane, also shows a marked circadian rhythm of toxicity in animals. A ten-minute exposure to standard doses kills 76 percent of the mice during their active period, but only 5 percent at another time.[7]

Vitamins are seldom considered poisonous; some are, however, in large doses. Niacin killed one-third of the animals tested at an active time, but two-thirds during the rest period.

Many drugs we use that affect the central nervous system have different powers at different times. A dose of amphetamine given rats at the end of their activity period killed only 6 percent of them, but the same dosage in the middle of the active time killed 78 percent.[8]

Doctors disregard human rhythms not only when they write out a prescription, but even earlier in a visit to their offices. Generally a visit to any medical facility, whether for a routine checkup or for treatment of a specific complaint, includes a check of the vital signs. Blood pressure, pulse, rate of respiration, and temperature are measured.

There is, of course, a circadian variation in body temperature. The other functions have cycles as well. A normal heartbeat can vary up to eight beats per minute over the course of a day, and the cardiac output, or the amount of blood the heart forces throughout the body, varies, reaching its maximum between ten in the morning and six at night.[9] The highest level of respiration comes in the early afternoon, and it is lowest in the early hours of sleep, when asthma usually attacks.

Blood pressure also follows a daily rhythm, with a peak between five and seven at night. For the normal person, the blood pressure varies about 10 percent over the course of the cycle; for the hypertensive the change can be even more substantial, with swings of up to forty points in the systolic and twenty points in the diastolic measurements. (Heart attacks, too, have predictable cycles. The most likely time for a heart attack is between eight and ten in the morning or evening.)

The great changes in our bodies are manifestations of innumerable others in our individual cells. The changes influence the structure and appearance of the cells. In the case of the liver, for instance, at one time of day it is storing glycogen, or sugar, in anticipation of the day's activity; at another time, after dispensing its fuel to provide energy, little remains.

Such activity is a necessary part of the body's cycle. But if a cancer is suspected to be growing in the liver, a biopsy must be performed. In order for the liver cells to be distinguishable under the microscope, it is necessary to dye them. As is to be expected, the time of day also influences how the cells react with the dye and how they look. The tissue samples can be incorrectly diagnosed if the circadian changes of the organ are not considered.[10]

For all that man has learned about the workings of his body and mind, there remain whole worlds to be uncovered. Mental illness is one area that is only partially understood, but research is emerging which shows that problems in the circadian system may play a role.

Howard Wilcox was a promising engineer in his early thirties who held a responsible position in a Los Angeles aerospace firm. He was required to travel frequently to the east coast. The number of such trips increased; he began to suffer more and more side effects, but in an unusual way. The trip east exhilarated and refreshed him, but the trips home were increasingly difficult. As time passed, he could hardly bear the thought of another return flight to the west and often required several days to recover before he returned to work. After one such trip, Howard Wilcox had a psychotic breakdown.

When he was tested, it was discovered that his natural rhythm was not twenty-five or even twenty-four hours, but twenty-three. He enjoyed his trips east because the time cues around him were brought more nearly in phase with his own shorter rhythm. When he traveled west, he was going away from rather than coming into his own rhythm. At the time of his breakdown, he could no longer adjust.

A number of psychotics have been reported to have rhythms differing from the twenty-four-hour period, even though they have the usual light and social time cues. In some cases of manic depression, for example, the rhythms run faster in the manic or active stage, and slower during depression, suggesting that circadian desynchronosis may be a factor in or at least a symptom of mental illness in certain patients.

There are also apparent seasonal rhythms in psychotic episodes, which some think may be caused by a primordial hormone reaction to photoperiodism, the change in the day length. A change of only a half hour of daylight can cause a major hormonal shift in the hamster, and melatonin, manufactured by the pineal gland under the stimulation of light, appears to be lower during depression and higher in mania. There is also an apparent shift in the daily cycle: the peak in melatonin in normal subjects came just before dawn, whereas in manic patients it came earlier in the night.

There is ample evidence that mental depression can result from a combination of factors, including adrenal gland activity, sleep rhythm problems, and the cycles of chemistry in our brain. In recent tests conducted at the National Institute of Mental Health, Drs. Frederick K. Goodwin and Thomas Wehr have produced a dramatic result in a manic-depressive patient in the depressed stage of the illness. By depriving the subject of one night's sleep and then moving the sleep schedule ahead a few hours, thus creating an accelerated schedule, the symptoms of the depression lifted for periods of up to two weeks. It appears that speeding up the slower rhythms of the depressed state thereby resynchronized the patient's rhythms, albeit temporarily.[11]

At the other end of the illness, at the manic stage when the rhythms are thought to be running at an accelerated pace, success has been reported in certain cases by using lithium. It is known that lithium slows or delays circadian rhythms in plants, and work by Dr. Daniel F. Kripke of the University of California at San Diego suggests that in certain manic-depressive patients with rapid cycles, lithium slows the circadian oscillator.[12]

One common sign that schizophrenics and manic depressives are entering a problem period and leaving one of more normal behavior is the abandonment of meat and vegetables and a ravenous craving for starch and sugar, a reflection of a change in chemical balances.[13]

A disturbance in the sleep rhythms of the mentally ill is a frequent signal of forthcoming episodes, from depression to psychosis.[14] The ubiquitous nature of wake-sleep disturbances in emotional illness may provide the missing link between all the theories and models explaining mental illness.

It has been established that desynchronosis of our rhythms leads to hostility. At the same time, psychiatric theory holds that unexpressed hostility brings depression. An unfortunate experiment suggests that one compounds the other.

Several college students volunteered for an experiment in shifting the circadian rhythm by changing the time cues. All were adjudged normal by the usual physical and psychological examinations. During the first week of entrainment, however, one of the volunteers showed a rhythm system out of synchrony with the others'. All the subjects were shifted to a schedule twelve hours different from local time and, after a few weeks, were returned to local time for ten days. They were then discharged.

At the end of the experiment, the volunteer who had had the aberrant rhythms during entrainment had not resynchronized, and his desynchronization had actually worsened. He became progressively more depressed and, two weeks after the conclusion of the study, he killed himself. In his case, the desynchronosis had led to depression, and the depression, in turn, to suicide.[15]

Mary Lamb, sister of nineteenth-century English essayist Charles Lamb, had a cycle of psychosis for over fifty years. She led a productive, normal life save for one dramatic, terrible episode: she killed her ailing mother. Her brother's method of dealing with her psychosis was to straitjacket her and deliver her to the hospital on the slightest sign of irritability. She spent most of her time living with her brother in the community and showed no sign of deterioration until the onset of senility in her later years.[16]

Time is a recurring theme in the study of mental illness. It rears its head again in the case of mental patients and their ability to calculate passing time. Asked to determine a certain time lapse, the patients' time sense is generally distorted; they underestimate the time passed, as if their rhythms, or nervous system, were running at an accelerated pace. Two characteristics of this confusion of time are revealing: the psychotic has a poorer sense of time estimation than the neurotic; and as a patient's condition improves, so does his ability to gauge time more accurately.[17]

The body's rhythms (and their disruption) may prove to be closely related to mental illness, but the relationship between circadian cycles and other illnesses has been clearly established. Sleep research in particular has revealed much about the mysterious hours our system is supposedly at rest, and this new information pertains to certain previously misunderstood or undiagnosed ailments.

Narcolepsy is an illness whose major symptom is a sudden paralysis. It can be partial or total, but it comes on instantaneously. Narcoleptics plod through life in a state of constant drowsiness occasionally broken by a full-fledged collapse into an apparent deep sleep; they return to consciousness immediately if they are touched by another person.

Despite the excessive sleepiness during the day and occasional naps, the narcoleptic sleeps a normal amount at night. His abnormal behavior occurs when he (or she) experiences feelings; most frequently laughter or anger, but virtually any excitement including elation, surprise, or the emotions aroused during lovemaking, or even something as simple as sighing brings about a reaction. The narcoleptic's response to such feelings may be as slight as a weakness in the knees or as debilitating as total paralysis.

Narcoleptics sometimes experience vivid dreams while they are awake, just prior to falling asleep, or upon awakening in the morning. This observation helps to portray narcolepsy for what it is: an unexplained breakdown in the mechanism that normally confines the REM patterns of dreaming and paralysis to sleep time. Most of the estimated 100,000 narcoleptics in the United States report having seen three to five doctors over ten to fifteen years, with no success, when the illness is finally diagnosed. This record should improve, and relief, if not cure, can be achieved through appropriate drugs.[18]

Another disease more thoroughly understood through sleep research is sleep apnea. It shares one symptom with narcolepsy, constant drowsiness, but the sleep apnea sufferer looks very different. He is cumbersome even in repose, his head nodding and his breathing heavy. Sleep apnea is the cessation of breathing during sleep. Until just a few years ago, those afflicted were a frustration to their doctors and a burden to their families. They dragged their way through life numbed by loss of sleep (without knowing it); the resultant fatigue made the slightest task an impossible chore, and the day was spent in high anxiety over the next night's sleep and, if they were aware of it, their high blood pressure.[19]

A person with sleep apnea can barely sleep; after a few minutes of repose and heavy breathing, and usually heavy snoring as well, the breathing suddenly stops as a result of the collapse of the airway, a malfunction of the brain, or both.[20] As the amount of oxygen in the blood drops, the mind arouses the body, and the huge, overweight mass jerks itself awake, gasping and heaving until the closed airway opens.

During such episodes, which can occur more than five hundred times a night, the person not only exerts a tremendous amount of physical effort but also awakens—although the victim remembers none of it. Having slept very little, he faces the morning more tired than when he retired.

One remarkable characteristic of the illness is that those afflicted with it live tortured existences for years, since the diagnosis has usually been reached when the patient is about fifty and the problem usually begins in the late twenties. Another amazing aspect of sleep apnea: it can be cured easily. A tracheostomy, the insertion of a small "pipe" into a small, permanent hole in the throat, is easily performed.[21] The fatigue and sleepiness disappear in a few days, the abnormal heart rhythms improve, the blood pressure returns to normal in two to three months, and weight loss is often dramatic.[22]

Impotence has no observable social symptom such as sleep apnea and narcolepsy, but, as has been said, the only thing worse than getting a social disease is not being able to get one. For years impotence was the domain of the psychiatrist rather than the physiologist, but new research from the understanding of sleep now points the way to the proper diagnosis and the correct treatment of the cause of impotence.

One can measure nocturnal penile tumescence (NPT), the quantity and the quality of erections that occur during the night. NPT begins not in preparation for puberty, but at three years of age, and has been found in subjects up to seventy-nine. The level of NPT rises to a maximum of over three hours a night in the prepubertal ages of ten to twelve, then gradually declines to about an hour and a half each night in a man's seventies. Again, man is not alone: NPT occurs in opossums, rhesus monkeys, and shrews; there is even an indication of a similar phenomenon in women.

NPT occurs in conjunction with REM sleep. Although NPT and REM usually occur concurrently, one does not cause the other; both are part of the sleep process, when the body and mind are geared to release instinctual energies. If REM sleep is suppressed, NPT is still experienced. Since REM episodes usually occur just prior to awakening in the morning, a man often awakes with an erection, a function of the REM-NPT process and not the result of a full bladder.

The amount of NPT during the night has no relation, contrary to myth or expectations, to recent sexual activities. While the suppression of REM does not dampen NPT, the dream content can influence it; if a dream is of an anxious sort, the amount of NPT during that episode decreases.

NPT is a measure of sexual function, and it should be used in the evaluation of every male who complains of impotence. In tests of over two thousand normal men with no sleep, drug, or alcohol problems, those with normal sexual function had normal NPT. Those complaining of impotence, however, had a high incidence of poor NPT. If the NPT patterns are abnormal, they point to a physical reason for the impotence, such as is frequently seen in diabetics; if the NPT is normal, it points toward psychological factors.[23]

Circadian science is opening new windows on the detection, prevention, and cure of cancer. As we have seen, the cells in virtually every tissue of our body divide or reproduce in a circadian rhythm. In cancer, however, the cells break out of their normal rhythmic behavior. They may divide faster or slower than they usually do, or there may be no rhythm at all.

Some researchers believe that people who are desynchronized or constantly consuming food and drugs that change the phase of the circadian rhythm may be a new high-risk group for contracting cancer. Coffee, tea, and certain barbiturates have been implicated as carcinogens or cocarcinogens; it is known that they shift the body temperature rhythm in rats. Our present urban societies are full of time cues as well as food and drugs that constantly rephase our rhythms. As an example of the relation of rhythms and cancer, it is known that an irregular menstrual cycle is one of the factors common in those who have a high risk of cancer of the breast, ovary, and uterus.[24]

Research from Europe indicates that it may be possible to predict breast cancer by detecting changes in the temperature rhythms of the breast. In many cases, women who show a change in the temperature rhythms of their breasts have no evidence whatsoever of any malignancy on physical examination. But a year or two later many return with a tumor. It is postulated that the temperature rhythm shifts before there is any structural change that can be detected. Hugh Simpson of the Glasgow Royal Infirmary has patented a brassiere that contains, without causing discomfort, detectors that indicate any shift in the temperature of the breasts.[25]

Some people die of cancer; others, unfortunately, die of the treatment—and some die of both. Because we are biochemically different at different phases of the circadian rhythm, we react differently to the identical stimulus or drug at different phases of the circadian cycle.[26] Research indicates there is an optimum time to treat various tumors with different drugs, and also with radiation. By doing so, scientists have been able to reduce the frequently very unpleasant side effects of the treatment and also to increase both the life-span of laboratory subjects and, in certain cases, the number who are completely cured (first human experiments are now being conducted).[27] Ignoring circadian rhythms can be dangerous in cancer research and treatment. Fortunately, a recent conference of cancer specialists outside the circadian field has recognized the need to include circadian considerations in both research and treatment.

Circadian rhythms are part and parcel of life, whether it be in sickness or in health. The further study of these rhythms can help increase the efficacy of the drugs we take and can help us understand more of the illnesses we experience, from cancer to the common cold.

He has called together legislative bodies at places unusual, uncomfortable, and distant from the depository of their Public Records, for the sole purpose of fatiguing them into compliance with his measures.
—The Declaration of Independence

8 Circadian Rhythms in Operation: Jet Lag and the Passenger

George III knew nothing of circadian science, but those who plotted his overthrow were well aware that some of the king's ploys—the scheduling of meetings at odd hours and at distant locales—were discomfiting indeed. The king's colonial opponents went so far as to cite such tricks in the Declaration of Independence.

The circadian disruptions they experienced were not unlike those that result from what the space age terms "jet lag." Present-day businessmen, politicians, and diplomats have been known to complain that their adversaries in negotiations try to make similar use of the effects of jet lag to take advantage of them. In the same vein, John Foster Dulles indicated in a deathbed interview with Marquis Childs that the decision he made as Secretary of State on the Aswan Dam "was one of the great mistakes of his life, and that he might have taken a more conciliatory stance with the Egyptians had he not been so weary from jet travel."[1]

Jet lag is the most obvious and most commonly felt disruption of circadian rhythms. In some instances it can cause a great deal more difficulty than the vague discomfort normally associated with it. An estimated 15 percent of those who pass through several time zones by jet experience no ill effects, but an equal number suffer severely. Those who get off lightly have a modest 8 percent loss of efficiency, but the 15 percent at the far end of the spectrum suffer a reduction of up to 70 percent in their levels of performance.[2]

Those who are familiar with aviation—airline medical directors, pilots, businessmen—describe the people who get off easily as having a quiet, laconic personality, the type who seem always to be able to fall asleep. Those who experience serious problems have personalities that are more rigid; they may be compulsive and usually exhibit a constant awareness of time.

Hubertus Strughold, a physician considered by many to be the father of space medicine, estimated in 1968 that the number of Americans who annually crossed four or more time zones was in the tens of millions.[3] Jet lag results from crossing only three, not four time zones, and since the volume of travel, particularly to Europe, has increased dramatically in the last decade, jet lag now affects about 25 million Americans each year. In short, it has become a malady that occurs more often than any except the common cold. (Only the Eskimos have no problem adjusting to time zone changes, a trait thought to be caused by their underdeveloped circadian rhythms.)

When man traveled by cart or train or early airplane, the change in time zones wasn't fast enough to cause the same discomfort jet lag does. (The complaints in the Declaration of Independence were caused by the odd, irregular hours similar to the problems encountered in shift work.) Only with the advent of the transcontinental and intercontinental jets that fly faster than the body and mind can adjust did jet lag per se come to be. (Ironically, there is some indication that supersonic planes reduce jet lag by being so fast that the time cues at the destination seem less out of sync with the body's clock. A morning flight from the east coast of the United States can deliver its passengers to London in time for dinner, which provides them with a social and mealtime cue and makes the resynchronization process easier.)

Jet lag is the result of a double-barreled dose of desynchronization, external and internal. Before taking off, the rhythms are synchronized: the heart, liver, kidney, adrenal glands, and other systems have the proper relationship to one another. But upon arrival at a destination several time zones away, the time cues are given and received at unexpected, unaccustomed times, and the various rhythms become uncoupled. Internal desynchronization results.

On the morning after flying from New York to London, the low point of body temperature is at ten in the morning rather than at five. (See Figure 2.) The plateau of peak performance does not come until mid- or late afternoon London time, rather than late morning. If a traveler goes from New York to Honolulu, the low point in his body temperature is at midnight. His performance plateau runs from five in the morning to mid-afternoon.

Although there is no recorded case of a traveler spontaneously perishing from jet lag, the symptoms can range from uncomfortable, to unpleasant, to serious. Simple irritability is a common problem, and there may be hostility and aggression, traits unsuited to negotiation or diplomacy. Physically, extreme fatigue is common, as are intestinal disturbances such as nausea or diarrhea. These symptoms, as well as the disruption of the

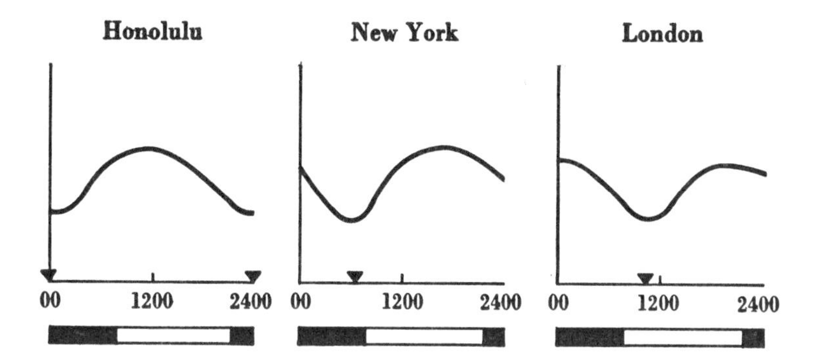

Figure 2. Jet Lag

The three charts show a New Yorker's circadian rhythm at home and in Honolulu and London on the first day of arrival. The local time and a typical wake (light)-sleep (dark) activity cycle are illustrated below each graph. The small triangle in each indicates the low point of deep body temperature.

Flying west to Honolulu, much of the peak performance time occurs during working hours, but in flying east to London, the performance plateau is largely outside working hours.

Eventually (one day for each hour of time zone change), the New Yorker's circadian rhythm will adjust so that he will be in synchronization with his new locale, as he was in New York; upon return to New York, he again will be out of synchronization.

Note that the amplitude or height of the curves in London and Honolulu is less when one is in desynchronization.

menstrual cycle, are immediately apparent. The most serious problems are those of which one may be unaware.

One activity that demonstrates the hidden symptoms is simply driving a car. Peripheral vision is reduced; response time to visual stimuli, such as turn indicators on another car or the change in a traffic signal, is also slowed; and night vision hindered. Reflexes too, are less effective, and the ability to adjust to fast focus (to glance from the speedometer to an oncoming vehicle, for example), is impaired.

For the business person, student, athlete, or anyone who

needs to perform upon arrival, there can be serious complications. Jet lag frequently brings faulty judgment and decreased motivation, or worse, a seeming stupidity. Jet lag can result in what is called "recent memory loss." The businessman may find himself asking for a sheaf of papers at one moment, then a few minutes later asking for them again. Time and place disorientation is common. Performance blocks appear, and responding normally to a ringing telephone or a knock at the door can seem like a confusing, arduous task.[4]

Probably the most common complaint, and a very serious one because it perpetuates the other problems, is the inability to sleep well.[5] The logical solution, it would seem, is to use an artificial sleep inducer such as alcohol or a sleeping pill. But, as in the case of other sleep disorders, these remedies actually hamper rather than promote proper sleep, since they do not deliver all the required types of sleep. After a night of heavy alcohol- or pill-induced sleep, one is less rested than after a sleep that seemed fitful.

Without proper rest, "mini-sleeps" can result. These are brief periods, only a few seconds long, when one actually falls asleep during the daytime, with eyes open. Mini-sleeps are, at times, embarrassing; if they occur at the wheel of a car, they can be fatal.[6]

Time is the cure for all such problems. After about *one day for each time zone crossed,* the wake-sleep cycle and some of the other rhythms are in external synchronization with the new place. Total internal resynchronization comes more slowly.[7] After a flight through twelve time zones, the kidney function, for example, requires much more time to synchronize; its rhythm is resynchronized only after some twenty-five days. The whole process is, of course, repeated after the return flight.

The effects of west-to-east and east-to-west travel are different. Although some people are exceptions to the norm, four different tests on transatlantic flights involving twenty-six people found that in every instance west-to-east travel was more difficult. There was greater desynchronization, and a longer period of adjustment was needed.[8] (See Figure 3.)

On reflection, the reason should be obvious: man's natural rhythm is twenty-five hours. On a flight east, the body's rhythm must be advanced or compressed, which is less agreeable than expanding the day. A perfect fit with the new surroundings does not occur, but at least the pain of the shorter, compressed time period or day is avoided. Some have attributed the difference to leaving or returning home. Although there may be a superficial euphoria one way or the other, the truth is not changed: the body and mind of most people have an easier time going west.

A north-south flight may cause some fatigue from other factors in the flight, but there is absolutely no evidence of circadian disruption. Jet lag does not result.

In preparing for his record-breaking around-the-world flight in 1933, Wiley Post was aware that on his previous trip around the earth he had suffered from "prop lag." To overcome this, he decided to reset his clock gradually to Moscow time, the halfway point. Before his departure, he rose and retired an hour earlier each day until he was synchronized with Moscow. Upon his return, Post claimed the system allowed him to make the flight without using stimulating drugs that other aviators of the time used on such flights.[9]

The two best preventives of jet lag are adjusting the internal clock to the schedule of the destination before leaving home or the reverse, staying on "home" time during the trip. The latter has some obvious practical limitations. Business in Europe cannot always be delayed until midafternoon, and a companion in Hawaii would probably find it strange to dine in the middle of the afternoon.

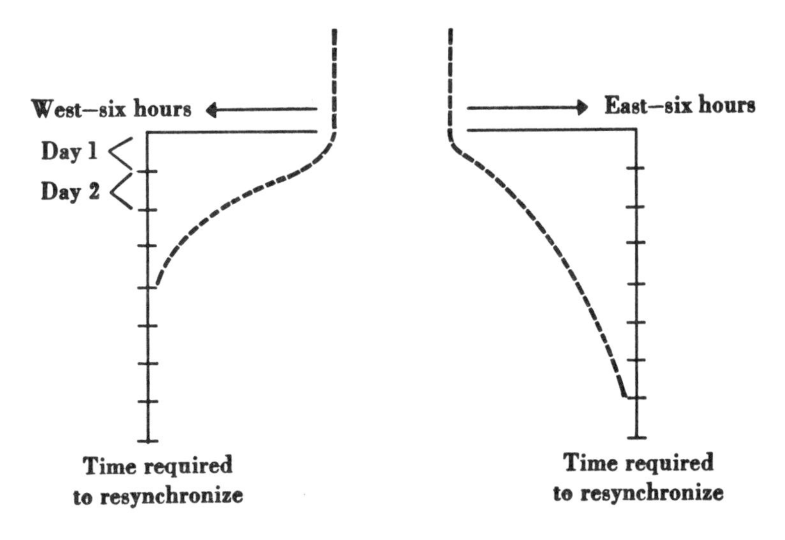

Figure 3. Flying East versus Flying West

This schematic drawing delineates the relative effects of flying east to west and west to east. Prior to departure, deep body temperature (the dotted line) is vertical, occurring at the same time each day. The vertical lines on the far left and right show the time when the low point of temperature should occur in the new time zone.

Flying six hours to the east causes deep body temperature to become more desynchronized, so a longer time is required for the body to readjust to the new environment.

Some *dos* and *don'ts* can facilitate the adjustment to the new time zone:

The simplest of the tricks one can use to minimize the effects of the circadian discrepancy is contained in a simple rhyme whose message should be followed for a few days after a flight: in the morning, when going east, do the least; in the morning, when going west, you're at your best.

Another point to remember: *do not* use either hard liquor or sleeping pills (unless you are using the latter under the care of a physician) immediately before, during the first few days of, or right after a trip. Smokers should try to reduce their consumption of tobacco. Both alcohol and tobacco deprive the body of oxygen; tobacco interferes with the transportation of oxygen in the blood, and alcohol blocks the cells' use of oxygen.

An oxygen lack is also a consequence of simply flying. People living at sea level may complain about the altitude in "mile-high" Denver, but on a transcontinental or overseas flight the passengers spend several hours in a cabin altitude far higher. The effects of alcohol and tobacco resemble those of altitude-induced oxygen starvation, and vice versa. As reported by Dr. Ross A. McFarland of the Harvard University School of Public Health, several cigarettes and two or three good drinks result in an effective "altitude" of twelve to fifteen thousand feet.[10]

Overeating, too, can worsen the effects of jet lag. *Do not* eat heavily for two or three days before flying three time zones or more. Avoid doing so during the flight and for a few days afterward. Heavy, rich, or unusual meals may often seem de rigueur on both business and pleasure trips, but they aggravate jet lag.

Dr. Charles Ehret of the Argonne National Laboratory, an expert on cell physiology and the metabolic pathways of animals, has developed a diet to counter jet lag. Using rats, Dr. Ehret simulated flights from Chicago to Europe and back again by changing the time givers in his laboratory. In both "directions," the rats who were fed the jet lag diet did not become internally desynchronized and adjusted to the new environment much more quickly. The diet is presently being tested by the army, which foresees a major jet lag problem should combat troops have to be moved quickly from the United States to other parts of the world. For those who wish to try it, the diet is to be found in Appendix C.[11]

Upon arrival, *do* jump into the habits or regimen of the new time zone as quickly as possible. A 9:00 A.M. flight from New

York to London arrives at 9:00 P.M. London time. Although it is only four in the afternoon back in New York, go to bed and try to sleep. Since aspirin lowers the body temperature, its ingestion may help resynchronize the rhythms, as well as help induce sleep, as mentioned in Chapter 4.[12] Although a jet from New York to Honolulu delivers its passengers in late afternoon, Hawaiian time, the New Yorker's metabolism thinks bedtime. Nevertheless, try to stay up for a few hours until the Hawaiians are retiring.

Do it together. The society of other people faced with desynchronization can help one adjust more rapidly. In one study, a group of travelers was divided into two sections. One group was restricted to the hotel and the company of their fellow travelers and was limited to meal and light changes as time givers. The other group was taken out of the hotel and involved in the life of the locale. Those who remained in the hotel adjusted at the rate of thirty-six minutes per day; the others resynchronized at the rate of ninety-two minutes per day, over two and a half times as fast.[13]

Do take modest exercise—a good walk, a brief jog, a swim. When flying east, exercise, then have a light supper with a moderate amount of wine, and go to bed. When going west, delay the exercise for a few hours, until late afternoon local time, and follow the same routine. Why a modicum of wine is helpful and hard liquor bad no one knows, but according to Charles Winget of NASA, wine enhances sleep without affecting its sequence of stages.

If you are on regular medication, *do* consult with a doctor as to the schedule in the new time zone. Doctors warn that diabetics taking insulin should be particularly careful.

Do plan layovers between the departure point and the destination on trips that cross more than six time zones.

Do arrive a day or more early for important conferences or events. It may seem like a waste of time and money, but the additional out-of-pocket cost is well worth the investment. One international financier who is in excellent physical condition and travels between California and Europe reports he is completely "washed out" for several days after a trip, particularly to the east, even though he usually lays over a day in New York to make plane connections. He wisely refuses to do any business of a serious nature at either end for three to four days.

There is no doubt in the minds of those medical researchers who have studied jet lag in Japan, Europe, and this country that travel across time zones brings a physiological debt that must be made up, even under the best conditions. There is also evidence that the debt may accumulate, as is illustrated by the case of Lowell Thomas. As reported in *Human Performance in the Aviation Environment,* "In 1963, he was hospitalized for a suspected myocardial infarction [heart attack]. When this was not confirmed, he realized that the better part of the past year had been spent chasing news stories in all parts of the world, constantly flying from place to place. He estimated that in the previous few months he had crossed all twenty-four time zones twice and sometimes more! His symptoms included muscular tremor, vertigo and syncope [fainting] and extreme fatigue."[14]

A similar case is that observed by a Pan American pilot who was offered a five-year contract by BOAC. He accepted and spent the time on routes between London, Paris, and Rome. After the contract expired, he returned to Pan Am and was shocked by the appearance of his former colleagues. He could easily divide them into two groups. One appeared to have aged normally; the other appeared older than the ages of its members warranted. It took him a few weeks to discover the explanation: those who seemed older had been flying east-west routes, while the pilots who appeared more typical of their age were flying north-south.

The effect on the aging process of constant phase shifting has not been tested with man, but experiments have been done on other organisms. A hatch of Mediterranean fruit flies was divided into two groups. One was placed in a constant light-dark cycle of 12:12. The other group was also given a 12:12 cycle of light and darkness, but every week the phase of the light was shifted by six hours, just as if the flies had traveled each week to a new home six time zones away. The flies with no shift in light-dark cycle lived for an average of 125 days; those who had been traveling about the earth lived for 98 days.[15] Similar experiments with mice, rats, and hamsters have given similar results: constantly changing time zones reduces the life-span.[16]

Human error was to blame for the crash of a military jet at a top secret U.S. base three months ago that killed all twenty men aboard, an Air Force accident report shows. It said a plane too heavy and a runway too short combined with a miscalculation by the crew caused the September 14 crash at a nuclear weapon storage facility. The report said that the crew of the EC-135, a Boeing 707 modified for use as an airborne command post, had worked for more than sixteen hours that day while participating in a secret practice mission. —*Los Angeles Times*, December 19, 1977

9

Circadian Rhythms in Operation: Jet Lag and the Pilot

Flying by jet involves highly sophisticated technology. Satellites and computers collect and analyze enormous quantities of data to determine and predict the weather en route. Expert mechanics scrutinize each aircraft for possible defects or signs of wear. The crew inhabits an "office" that is, if not as commodious, certainly far more expensive than that of a corporate president. Most routes have been traveled hundreds and hundreds of times. Given all this, what could possibly go wrong? Perhaps a few anecdotes will elucidate the problem.*

* The issues of airline safety and the rules and regulations governing the hours a cockpit crew may work, and even the definition of work or duty, are extremely sensitive ones. To secure maximum cooperation, all interviews in this chapter were conducted with the understanding that the parties would not be identified.

• A transcontinental flight was proceeding smoothly; and the crew put the plane on automatic pilot and relaxed for a few minutes before preparing to land in Los Angeles—standard operating procedure. But then all three crewmen went to sleep and overflew their destination by a hundred miles before being awakened by the stewardesses.

• Another crew, exhausted by a long international flight, landed in Chicago and, while waiting for clearance to go to the gate, fell asleep.

• Another perfect landing was made in Sacramento and, as the plane decelerated, the pilot, in his early thirties, died of a heart attack.

• Approaching Denver after a tiring tour of duty, another pilot fell asleep for a few seconds while the plane was on approach. He awakened to find the plane perfectly aligned with the taxiway, not the runway.

Such incidents usually remain the dark secrets of aviation; they also point to one of the flaws in the otherwise enviable record of American aviation: not enough attention is paid to the human factor in flying. New research and knowledge relevant to a person's ability to perform at a certain time is seldom heeded.

If there were no human factor involved, pilots would not be needed for today's sophisticated aircraft. Such is not the case. In fact, the aircraft is an extension of the pilot's brain and body controlled by his hands and feet. He belts himself to two hundred or more tons of equipment. He is aided by a crew and enough electronic gadgetry to fascinate Thomas Edison for years. Yet the human factor in flying has not been eliminated. Studies made by Dr. John Lauber of NASA indicate that on one out of every five flights the pilot makes a mistake serious enough to cause an accident.[1]

The checks and double-checks, backup systems, and alternate procedures used, and the millions of dollars spent by the airlines and aircraft manufacturers in the design and development and maintenance of planes, prevent more accidents and fatalities from occurring. The industry can point to millions of passenger miles flown without a mishap, to the fact that, statistically, flying is safer than driving a car. But when an accident does occur, it can be devastating.

On April 22, 1974, a Pan American 707 crashed in Bali, killing all aboard—ninety-six passengers and the crew of eleven. A party intimately acquainted with the accident states that only one thing—fatigue—could account for the "collective stupidity," misreading a radio direction beacon, of the experienced crew. And a look at the schedule of the crew, shown on their home base, San Francisco time, explains why:

	San Francisco	
	Day	*Time*
Depart San Francisco	17	7:44 P.M.
Arrive Honolulu	18	1:32 A.M.
Depart Honolulu	19	3:39 A.M.
Arrive Sydney	19	2:35 P.M.
Depart Sydney	20	6:21 P.M.
Arrive Jakarta	21	1:30 A.M.
Depart Jakarta	21	2:18 A.M.
Arrive Hong Kong	21	6:40 A.M.
Depart Hong Kong	22	4:00 A.M.
Crash Bali	22	8:30 A.M. (approx.)

It would be difficult for someone staying at home to follow this work schedule and perform even routine tasks particularly well. After the crash the schedule was altered.

On September 25, 1978, a PSA jet liner crashed in San Diego, killing all aboard, the worst domestic air disaster in our history. Although the official report of the accident has not been released at this time, the schedule of the crew just prior to the accident is known:

September 22	Three flights beginning at 4:30 P.M. and ending at 11:50 P.M.
September 23	Two flights beginning at 6:45 A.M. and ending at 9:40 A.M.
September 24	Four flights beginning at 4:40 P.M. and ending in Sacramento at 10:38 P.M.
September 25	Depart Sacramento at 7:00 A.M., stopover in Los Angeles, crash in San Diego approximately 9:00 A.M.

On two of the three nights prior to the crash—September 22 and 24—the crew had had abbreviated periods of rest, probably less than five hours on the twenty-second and less than six hours on the twenty-fourth. In addition, on the twenty-second they had worked a "swing" shift, on the twenty-third a morning shift, on the twenty-fourth another swing shift; and on the day of the crash, they were back to a morning shift.

There is not a simple answer to the question "Why?" in the case of these crashes. But to begin with, the problem of jet lag is not confined to the passengers, nor even to the "mid-continent man" described in Chapter 6. Pilots and other airline personnel suffer similar and even compounded effects. Yet at the time of this writing, the Federal Aviation Administration and the airlines themselves have demonstrated no inclination to consider circadian desynchronization and its potential hazards in setting their rules, regulations, or schedules.

One major airline flies a potentially hazardous pair of flights for the crew, and therefore the passengers, between California and Hawaii. The first leaves San Francisco at nine in the morning and arrives in Honolulu at two in the afternoon, California time. The crew then has a twelve-hour layover, during which they are required to get at least six hours of rest and eat a meal. But sleeping during the allotted time period is not easy: the pilot and his crew are being asked to slumber when their minds and bodies are accustomed to being awake.

The return flight departs Honolulu at 2:00 A.M. California time and arrives in San Francisco shortly before seven in the morning. The pilot and crew fly a total of ten hours out of twenty-two, from 9:00 A.M. to 2:00 P.M. and from 2:00 A.M. to 7:00 A.M. They must also check out and prepare for each of the flights. Most important of all, they fly the return leg, often without the required six hours' rest, when their circadian rhythms are at ground zero. This California-Hawaii-California tandem breaks no rules; it is completely within the regulations established by the Federal Aviation Administration and the contract between the pilots and the particular airline involved.

Karl E. Klein of the DFVLR Institute of Aviation Medicine in Bad Godesberg, Germany, has pointed out what should be obvious: night flights should be undertaken only by crews who are well rested.[2] When people are in the sleep mode, they have reduced readiness for mental and physical performance. Clearly, people should not be required to display maximum efficiency when they are at the low point of their daily rhythms, and particularly people performing duties whose precise execution is so critical to the lives of so many.

Those who defend this schedule point to the fact that there

has never been an accident on this series of flights. But, for instance, would it be in the best interests of a medical patient to have a surgeon and surgical team conduct an operation under a similar schedule? The obvious answer is that at times there are emergencies when it is necessary, and successful, but that does not mean that such a practice is desirable. As Sir Peter Masefield stated in the foreword to *Pilot Error,* "If an accident *can* happen, sooner or later it will happen."[3]

When queried about the importance of circadian rhythms in flying, the medical director of the airline in question responded that he knew that some flights were more difficult than others, but that the pilots were not overworked; it is their responsibility, he said, to deliver themselves well prepared to execute the schedule within the government regulations. He acknowledged that the FAA had done some research on circadian rhythms—but he did not know precisely how the test data applied and concluded that the tests mean so much and yet so little. When specifically asked about Klein's statement regarding performance at the low point of the circadian cycle, he denied that safety was affected, stating that adrenaline may compensate for a pilot's being at the low point of the rhythm.

Physicians and researchers express disbelief when asked if adrenaline can be used to compensate for being at the low point of the circadian rhythm. Dr. Stanley R. Mohler, an expert in circadian rhythms formerly affiliated with the Office of Aviation Medicine of the FAA, stated in an interview that fatigued pilots, despite adrenaline, perform less well. Adrenaline secretion is not a process that can be turned on and off at will, and it may respond to stress too soon or too late, particularly with overuse or aging.

Despite evidence from physicians and even government

agencies about the dangers in flaunting the circadian cycles, the medical directors of airlines remain adamant that there is no problem. The medical director of one major airline does not believe the research valid since it had, in his opinion, no relation to the real situation. A pilot just before the end of a long transmeridian flight, he said, might not perform well on tests but, when nearing the destination, would pull himself together. He would then do very well on the approach and landing and would do well on a test given at that time. A half hour later, the fatigue would take over and he might well perform poorly on a test.

Researchers, however, think their findings are valid and do bear on the real situation. They are unconvinced that people who are "off" can suddenly turn on; in particular, they question whether it can be done time after time.

They point out that the physical stress a plane can bear is tested in the laboratory. It should be the same for human factors. Klein and his group of researchers have flatly stated that long transmeridian flights should be restricted to pilots no older than forty-five to fifty years of age. Many of the Boeing 707s, having been in service twenty years, have been reinforced. Through the consideration of their circadian rhythms, pilots could be given a dissimilar but equally important advantage.

Although some pilots do not object to such scheduling as that between San Francisco and Hawaii—a number because they do not suffer from jet lag and others because they make more money on such flights—many do complain. They are naturally and properly concerned for their own well-being, their ability to continue to pursue their chosen career and even to live long enough to enjoy their years of retirement. Beyond this is a feeling of professional responsibility. The concern is not that they do not feel well, but the recognition that they are not performing as well as they should. As one pilot on the San Francisco–Hawaii run rather bluntly expressed it, "I feel shitty and I'm doing a shitty job."

A pilot in the 1970s is not so much a mechanical, physical operator of the aircraft as he is a monitor. He must be sensitive to the minute variations in the many complex systems. Three decades ago, accident reports usually concerned mechanical failure. Today, with the advances in jet aircraft and technology, over half the accidents, 55 percent, are caused by pilot error.[4]

On October 13, 1976, an air freighter crashed in Santa Cruz, Bolivia, killing all three crew members and seventy-seven people on the ground, with seventy-eight others seriously injured and eleven missing. The plane had left Houston carrying drilling equipment the afternoon of the twelfth and had flown all night, with a stop in Curaçao before arriving at Santa Cruz at six in the morning. The crew then retired to a nearby hotel but returned to the airport in time to prepare for a takeoff at approximately 12:30 P.M.

The craft taxied down the runway, then became airborne. It failed to gain the necessary height to clear a tree and reinforced concrete pole located about 250 yards from the end of the runway. The plane then struck other obstacles and disintegrated, spewing debris over a populated area and subsequently crashing upside down. At the time of the incident, the crew had been in service 23.41 hours. Three witnesses who observed the crew during their layover in Santa Cruz reported they appeared to be "very tired."[5]

Though an FAA inspection in Houston cited some problems with the condition of the aircraft, the accident report stated, "From the physical evidence found at the time of the accident, explosion or fire aboard, structural failure, malfunction of flight controls, as well as systems failure are ruled out as probable causes of this accident." The accident report also stated, re-portedly for the first time ever in a major airline accident, that crew fatigue was a contributing cause:

> It is concluded that the probable cause of the accident was due to a possible error on the part of the crew in not using appropriate thrust setting to obtain the necessary acceleration, followed furthermore by a premature take-off, with the result that they obtained a reduced acceleration and climb capacity, reaching a very low altitude to be able to clear the obstacles on the take-off trajectory.
>
> The premature turning and take-off of the aircraft probably was the result of a natural impulse of the crew; seeing the imminent end of the runway, to be able thus to clear the obstacles on the take-off trajectory. The factors that may have contributed to the accident include: a) crew fatigue, added to the effects of high air temperature in the cockpit; b) crew error in not recognizing in time their low acceleration, to abort the take-off; c) operator's error in not requiring the completion of the rest period established for crew members, in conformity with the regulations in force.[6]

If pilot error is the cause for over half our airline accidents, there are only two possible explanations: either the pilots are unqualified, or there is a reason for qualified pilots to make mistakes. No one believes the former explanation. And when pilots and others are willing to grant an error may have been committed, they usually provide several causes for the error. Most often, pilot and/or crew fatigue is cited.

Fatigue is difficult to define and virtually impossible to calibrate through medical tests and measurements.[7] The only real "proof" of fatigue is an admission by an individual that he is tired or tests that show his performance is slower and less reliable than in an unfatigued person. And this evidence perishes with the crew in a crash. Other pilots are loath to admit fatigue over their radios; it would be picked up by the voice recorder, hurting their careers.

Nevertheless, pilots and others associated with aviation are convinced that fatigue is the major reason for pilot error. The major causes of fatigue are stress and the disruption of the circadian rhythms; multiple landings and takeoffs in the same day (when the heartbeats per minute can rise to over 100 on takeoff and up to 130 on landing); loss of sleep or disruption of the sleep cycle; multiple night flights; and other disruptions caused by flying through several time zones.

As Kenneth Clark pointed out in *Civilization,* modern man has been living in an era of engineering and mathematics; as Lionel Tiger suggests, we are now entering the era of biology. This transition should be reflected in attitudes toward airline safety. Immense strides have been made in the refinement and improvement of the equipment, and now it is time to turn our attention to the biological factors in flying.

The problems with airline safety occur not only in the air but on the ground as well. As discussed in Chapter 6, the air traffic controllers, the people who regulate the takeoffs and landings as well as the positions of the planes in the air space, are cited by experts in the fields of both stress and circadian rhythms as potential trouble.

Every airport negotiates shift schedules with the local of the controllers' union, but controllers frequently shift their work schedules. They may work two or three days on one shift, then shift again for two days; they may work two shifts for two days and one shift for one day that week. The controller handling an air freighter that crashed in Utah in 1977 re-portedly had worked from 7:00 A.M. to 3:00 P.M. the day before, then the eleven-to-seven shift the night of the crash.

The physiological debt incurred by constant shifting is shown by the rate of retirement: on the average, controllers last only fifteen years on the job. Most flunk their physicals before reaching age forty, with ulcers, high blood pressure, irregularities in the patterns of their brain waves shown on the EEG, and psychological stress revealed by hand tremors.

Nevil Shute, a novelist and an aviation engineer, has said, "Experience has taught me one sad fact—that you can't sell safety. Everyone pays lip service to the safety of aeroplanes, but no one is prepared to pay anything for it."[8] Certainly everything that is economically feasible should be done to improve airline safety.

The further elucidation of kinetic mechanisms . . . should be a desirable objective for students of the behavior of living organisms, whether these students call themselves biochemists, biophysicists, physiologists, botanists, zoologists, behaviorists, or psychologists.

—Hudson Hoagland

10

Your Circadian Rhythms

Truly, our environment pulses with rhythms. As the sun rises and sets, all life on earth ebbs and flows with a constancy that matches the movements of the tides. Our cohabitants on earth, whether flora or fauna, march to their own drummers. And most of all, man in his ultimate complexity is a rhythmic creature.

Each of us has within us a fortune teller. As we go through the day, we are usually unaware of our changing moods and abilities, the rise and fall of our senses and needs. Yet at a given time today, we are what we were yesterday at this time, and at this time tomorrow we will again be the same.

Our bodies and minds love repetition. It provides a frame of reference. Our thoughts and actions are accustomed to the cues of the light-dark cycle, of our work time, mealtimes, our schedule of exercise and play, and a multiplicity of other timely events. The routine can be broken (indeed, it probably should be, if only to relieve boredom), but it should be our home base.

We have seen how circadian science can help us to perform

better. If we engage in certain activities when our bodies and minds are most predisposed to them, we can maximize our efficiency. If we follow the dos and don'ts rhythms research has suggested, we can adjust to changes in our schedules, or simply to our peculiarities.

An exquisitely simple principle is to be learned from circadian science: to paraphrase Socrates, "Know thine own rhythms." Health, wealth, wisdom, and even longevity can result from living in harmony with the ticking of our internal clock, from listening to and obeying it.

The benefits of circadian science are by no means limited to our individual existences. If pressure can be brought to bear on such governmental agencies as the FAA and the FDA, changes for the better can be made in their workings. Flight regulations should be changed. The crews are fatigued, and the forty-year-old regulations are wrong and outdated.

The Food and Drug Administration's rules regarding drug testing and research must be changed also. At present, the FDA issues a book entitled *Good Laboratory Animal Practices* for those involved in drug research. It does not contain a single statement about circadian rhythms. As a consequence, research laboratories of major pharmaceutical companies throughout the country keep test animals in constant or irregular light. The circadian systems of the animals are not functioning properly, and they are not suitable for proper drug testing.

Government intervention or, perhaps, an enlightened, open-minded initiative in the private sector can help to enhance productivity through knowledge and application of circadian science. Simply by reducing the number of shift changes—that is, by leaving workers on the same shift for three or four weeks or more—the workers will perform better and have fewer illnesses, and absenteeism will decrease.

And flexitime, too, can increase productivity and job satisfaction.

If we cannot rely on the government to foster the proper programs (at least until greater pressure has been put on our elected representatives by the general public as well as by the scientists and researchers), we can rely on ourselves. One thing that can—and should—be taught to every high school student is autorhythmometry, or the taking of one's own rhythms.

Autorhythmometry is easy to learn and is easily applied to other members of the family. Any significant change in body rhythms can be an early warning of disease. All the tests—pulse, blood pressure, estimation of time, body temperature, finger counting, adding speed, hand-grip strength, and expiration from the lungs—can be done in fifteen minutes or less. The most difficult of them, taking the blood pressure, is a simple procedure and can be mastered after a few tries. And if widely adopted, autorhythmometry can have widespread beneficial effects.

We should embrace our rhythms, use them to help us accomplish more and stay in better health. We should treat them with respect. In a variety of organisms, alcohol shifts the rhythms, and so do coffee, tea, and phenobarbital. This is not a call for abstinence, but for a regulation of our schedules and a consideration of the effects on our rhythms of what we introduce into our bodies.

Our properly functioning circadian system maintains the integrity of our entire being. Organisms with their rhythm systems intact resist strains placed upon them far better than those whose rhythms are in disarray. It is our internal environment, the synchrony and harmony of our circadian system, that determines what we are and what we will become.

Most people know that Louis Pasteur discovered bacteria

and developed the pasteurization of milk, but few outside medicine are familiar with his contemporary, Claude Bernard, the acknowledged father of physiology. During their concurrent careers, Bernard and Pasteur engaged in a relentless debate over which was more important to the health and harmony of the organism, Pasteur's discrete invaders from the outside or Bernard's concept of the proper functioning of the system within.

On his deathbed Pasteur relented. "Bernard was right," he said. "The terrain is everything."

Appendix A

Association of Sleep Disorders Centers

This is a list of members of the Association of Sleep Disorders Centers as of November 16, 1978. Please note that this is not an exhaustive list of institutions where sleep disorders medicine is being done. Furthermore, please note that centers have various procedures for accepting patients. Some accept patients only through referral from a primary care physician; others will see patients on a self-referral basis.

Sleep Disorders Center (999) 999-9999
Dept. Intern Geneeskunde
2610 Wilrijk
Antwerpen, BELGIUM
Attn: Olga Petre-Quadens, M.D.

Sleep Disorders Center (301) 396-5859
Baltimore City Hospital
Baltimore, MD 21224
Attn: Richard Allen, M.D.

Sleep Disorders Clinic (617) 734-6000
Boston Children's Hospital
300 Longwood Avenue
Boston, MA 02115
Attn: Myron Belfer, M.D.

Sleep-Wake Disorders Unit* (212) 920-4841
Montefiore Hospital
111 E. 210th Street
Bronx, NY 10467
Attn: Charles Pollak, M.D.

Sleep Disorders Center (215) 874-1184
Department of Neurology
Crozer Chester Medical Center
Chester, PA 19013
Attn: Calvin Stafford, M.D.

Sleep Disorders Center (312) 942-5000
Rush-Presbyterian-St. Luke's
1753 W. Congress Parkway
Chicago, IL 60612
Attn: Rosalind Cartwright, Ph.D.

Sleep Disorders Center (312) 649-8649
Suite 214 Wesley Pavilion
Northwestern University Medical Center
Chicago, IL 60611
Attn: John Cayaffa, M.D.

Sleep Disorders Center* (513) 861-3100
Cincinnati General Hospital
Cincinnati, OH 45267
Attn: Milton Kramer, M.D.

Sleep Disorders Center (216) 368-7000
Psychiatry Department
St. Luke's Hospital
Cleveland, OH 44118
Attn: Joel Steinberg, M.D.

* Fully accredited centers

Sleep Disorders Center (216) 795-6000, X531
Mt. Sinai Hospital
University Circle
Cleveland, OH 44106
Attn: Herbert Weiss, M.D.

Sleep Clinic* (614) 422-5982
Department of Psychiatry
Ohio State University
Columbus, OH 43210
Attn: Helmut Schmidt, M.D.

Sleep Disorders Center (313) 876-2233
Henry Ford Hospital
2799 W. Grand Blvd.
Detroit, MI 48202
Attn: Thomas Roth, Ph.D.

Sleep Disorders Clinic (603) 646-2213
Department of Psychiatry
Dartmouth Medical School
Hanover, NH 03755
Attn: Peter Hauri, Ph.D.

Sleep Clinic* (713) 790-4886
Baylor College of Medicine
Houston, TX 77030
Attn: Ismet Karacan, M.D.

Sleep Laboratory (501) 661-5272
Department of Anatomy
University of Arkansas Medical Center
Little Rock, AR 72201
Attn: Edgar Lucas, Ph.D.

BMH Sleep Disorders Center (901) 522-5651
Baptist Memorial Hospital
Memphis, TN 38146
Attn: Helio Lemmi, M.D.

Sleep Disorders Center (305) 674-2385
Mt. Sinai Medical Center
4300 Alton Road
Miami Beach, FL 33140
Attn: Marvin Sackner, M.D.

Sleep Disorders Center (612) 347-2121
Neurology Department
Hennepin County Medical Center
Minneapolis, MN 55415
Attn: Milton Ettinger, M.D.

Sleep Disorders Clinic (514) 333-2070
Hopital du Sacre-Coeur
5400 ouest, Boulevard Gouin
Montreal, Qu., CANADA H4J 1C5
Attn: Jacques Montplaisir, M.D.

Sleep Disorders Center (201) 456-4300
Medical Sciences Building
New Jersey Medical School
Newark, NJ 07103
Attn: James Minard, Ph.D.

Sleep Disorders Center (504) 588-5236
Psychiatry and Neurology Dept.
Tulane Medical School
New Orleans, LA 70118
Attn: John Goethe, M.D.

Sleep Disorders Center (405) 272-9876
Presbyterian Hospital
Oklahoma City, OK 73104
Attn: William Orr, Ph.D.

Sleep Disorders Center (714) 634-5777
U.C. Irvine Medical Center
101 City Drive South
Orange, CA 92688
Attn: Jon Sassin, M.D.

Sleep Disorders Center (613) 231-4738
Ottawa General Hospital
43 Bruyere
Ottawa, CANADA K1N 4C8
Attn: Roger Broughton, M.D.

Sleep Disorders Center* (412) 624-2246
Western Psychiatric Institute
3811 O'Hara Street
Pittsburgh, PA 15261
Attn: David Kupfer, M.D.

Sleep Disorders Center (213) 451-3270
1260 15th Street, Suite 1402
Santa Monica, CA 90404
Attn: John Beck, M.D.

Sleep Disorders Program* (415) 497-7458
Stanford University Medical Center
Stanford, CA 94305
Attn: Laughton Miles, M.D.

Sleep Laboratory (516) 444-2069
Department of Psychiatry
SUNY at Stony Brook
Stony Brook, NY 11794
Attn: Merrill M. Mitler, Ph.D.

Sleep Laboratory (617) 856-3081
Department of Neurology
University of Massachusetts Medical Center
Worcester, MA 01605
Attn: Sheldon Kapen, M.D.

Appendix B

*The Social Readjustment Rating Scale**

	Life Event	*Mean Value*
1.	Death of spouse	100
2.	Divorce	73
3.	Marital separation from mate	65
4.	Detention in jail or other institution	63
5.	Death of a close family member	63
6.	Major personal injury or illness	53
7.	Marriage	50
8.	Being fired at work	47
9.	Marital reconciliation with mate	45
10.	Retirement from work	45
11.	Major change in the health or behavior of a family member	44

*Reprinted with permission from *Journal of Psychosomatic Research,* Vol. 11, p. 216, T. H. Holmes and R. H. Rahe, 'The Social Readjustment Rating Scale," 1967, Pergamon Press Ltd.

Life Event	Mean Value
12. Pregnancy	40
13. Sexual difficulties	39
14. Gaining a new family member (e.g., through birth, adoption, oldster moving in, etc.)	39
15. Major business readjustment (e.g., merger, reorganization, bankruptcy, etc.)	39
16. Major change in financial state (e.g., a lot worse off or a lot better off than usual)	38
17. Death of a close friend	37
18. Changing to a different line of work	36
19. Major change in the number of arguments with spouse (e.g., either a lot more or a lot less than usual regarding child rearing, personal habits, etc.)	35
20. Taking on a mortgage greater than $10,000 (e.g., purchasing a home, business, etc.)	31
21. Foreclosure on a mortgage or loan	30
22. Major change in responsibilities at work (e.g., promotion, demotion, lateral transfer)	29
23. Son or daughter leaving home (e.g., marriage, attending college, etc.)	29
24. In-law troubles	29
25. Outstanding personal achievement	28
26. Wife beginning or ceasing work outside the home	26
27. Beginning or ceasing formal schooling	26
28. Major change in living conditions (e.g., building a new home, remodeling, deterioration of home or neighborhood)	25
29. Revision of personal habits (dress, manners, associations, etc.)	24
30. Troubles with the boss	23
31. Major change in working hours or conditions	20
32. Change in residence	20

Life Event	*Mean Value*
33. Changing to a new school	20
34. Major change in usual type and/or amount of recreation	19
35. Major change in church activities (e.g., a lot more or a lot less than usual)	19
36. Major change in social activities (e.g., clubs, dancing, movies, visiting, etc.)	18
37. Taking on a mortgage or loan less than $10,000 (e.g., purchasing a car, TV, freezer, etc.)	17
38. Major change in sleeping habits (a lot more or a lot less sleep, or change in part of day when asleep)	16
39. Major change in number of family get-togethers (e.g., a lot more or a lot less than usual)	15
40. Major change in eating habits (a lot more or a lot less food intake, or very different meal hours or surroundings)	15
41. Vacation	13
42. Christmas	12
43. Minor violations of the law (e.g., traffic tickets, jaywalking, disturbing the peace, etc.)	11

Appendix C

Jet Lag Diet

Eastbound Transatlantic—Departure Time 2130

Sunday	*Monday*	*Tuesday*	*Wednesday*	*Thursday*	*Friday*	*Saturday*
1	2	3	4	5	6	7
Feast	Fast	Feast	Fast	Break the fast with breakfast on Paris time + large lunch + large supper	Three full meals Paris time	Etc.

Eastbound **Day of Flight** **Wed** 4

avoid carbohydrates
fruit is OK

No Coffee
No Tea
unless decaff.

Eat
Sparingly
Fast
All Day

decaff. OK

no snacks
no cream
no sugar

Several cups
of
black coffee
or
strong tea

Lights Off
No Movie
Dark—Quiet—Rest

Day of Arrival **Thu** 5

Lights on
Big High Protein
Breakfast

no coffee
no tea
unless decaff.

0130 CST
0730 Paris
Time

Big
Lunch

Big
Supper

CST	0600	1200	1800	2400	0600	1200
Paris Time	1200	1800	2400	0600	1200	1800

Notes

Chapter One

1. Jean Jacques d'Ortous de Mairan, "Observation botanique," in *Histoire de l'Académie Royale des Sciences* (Paris, 1729), p. 35.
2. Henri-Louis Duhamel du Monceau, *La Physique des Arbres*, vol. 2 (Paris: H. L. Guerin and L. F. Delatour, 1758).
3. Augustin Pyramus de Candolle, *Physiologie Végétale* (Paris: Béchet jeune, 1832).
4. Alfred J. Ewert, *Pfeffer's Physiology of Plants*, vol. 3, ed. and trans. Wilhelm F. P. Pfeffer (Oxford: Oxford University Press, 1905).
5. W. W. Garner and H. A. Allard, "Flowering and Fruiting of Plants as Controlled by the Length of the Day," *Yearbook of the Department of Agriculture* (Washington, D.C.: U.S. Government Printing Office, 1920), pp. 377–400.
6. Richard J. Wurtman, "Effects of Light and Visual Stimuli on Endocrine Function," *Neuroendocrinology* 2 (1967).
7. Stephen Jay Gould, "The 120-Year Bamboo Clock," *Natural History*, April 1, 1977, p. 8.
8. Gay Gaer Luce, *Biological Rhythms in Psychiatry and Medicine* (Washington, D.C.: National Institute of Mental Health, U.S. Department of Health, Education and Welfare, 1970), p. 8.

9. "New Facts on Biorhythms," *Science Digest,* May 1976, p. 70.
10. Ibid.
11. Ibid.

Chapter Two

1. David S. Saunders, *An Introduction to Biological Rhythms* (New York: John Wiley and Sons, 1977), p. 70.
2. Ibid.
3. Ibid.
4. Klaus Schmidt-Koenig, *Migration and Homing in Animals* (Berlin–Heidelberg–New York: Springer-Verlag, 1975), pp. 20–21.
5. Ibid., p. 61.
6. Ibid., p. 55.
7. Ibid., pp. 24–25.
8. Ibid., p. 56.
9. Saunders, *Introduction to Biological Rhythms,* p. 70.
10. John D. Palmer, *Introduction to Biological Rhythms* (New York: Academic Press, Inc., 1976), p. 206.
11. Garner and Allard, "Flowering and Fruiting of Plants."
12. J. N. Mills, ed., *Biological Aspects of Circadian Rhythms* (London and New York: Plenum Press, 1973), p. 281.
13. Luce, *Biological Rhythms,* p. 11.
14. Palmer, *Introduction to Biological Rhythms,* p. 75.
15. Saunders, *Introduction to Biological Rhythms,* p. 78.
16. Benjamin Rusak and Irving Zucker, "Biological Rhythms and Animal Behavior," in *Annual Review of Psychology* (Palo Alto, Calif.: Annual Reviews, Inc., 1975).
17. Palmer, *Introduction to Biological Rhythms,* pp. 26–27.
18. Ibid., p. 184.
19. Ibid., p. 76.
20. Ibid., p. 86.
21. Saunders, *Introduction to Biological Rhythms,* p. 150.
22. Ibid., p. 154.
23. Janet Raloff, "Biological Clocks, How They Affect Your Health," *Scientific Digest,* November 1975, pp. 65–66.
24. Saunders, *Introduction to Biological Rhythms,* p. 142.

25. Ibid., pp. 134–136.
26. *Newsweek,* May 1, 1978, p. 80.

Chapter Three

1. Colin S. Pittendrigh, Director of the Hopkins Marine Station of Stanford University, Stanford Alumni Lectures, Los Angeles, Calif., February 1977.
2. Luce, *Biological Rhythms,* p. 121.
3. Ibid., p. 14.
4. J. Aschoff, M. Fatranska, H. Giedke, P. Doerr, D. Stamm, and H. Wisser, "Human Circadian Rhythms in Continuous Darkness: Entrainment by Social Cues," *Science* 171 (1971).
5. Workshop on "Biological Clocks," Hopkins Marine Station of Stanford University, Pacific Grove, Calif., Summer 1977.
6. Ibid.
7. Ibid.
8. Ibid.
9. R. T. W. L. Conroy and J. N. Mills, *Human Circadian Rhythms* (London: J. & A. Churchill, 1970), p. 25.
10. Workshop on "Biological Clocks," Stanford University; William C. Dement, *Some Must Watch While Some Must Sleep* (San Francisco: San Francisco Book Company, Inc., 1976), p. 108.
11. Jürgen Aschoff and Rütger Wever, "Human Circadian Rhythms: A Multioscillator System," *Federation Proceedings* 35 (1976): 2326–2332.
12. Palmer, *Introduction to Biological Rhythms,* p. 151.
13. Workshop on "Biological Clocks," Stanford University.
14. Ibid.
15. Ibid.
16. Aschoff and Wever, "Human Circadian Rhythms."
17. Luce, *Biological Rhythms,* p. 39.
18. Workshop on "Biological Clocks," Stanford University.
19. J. Aschoff and H. Pohl, "Rhythmic Variations in Energy Metabolism," *Federation Proceedings* 29 (1970): 1541–1552.
20. Conroy and Mills, *Human Circadian Rhythms,* pp. 102 ff.
21. Ibid., p. 100.

22. J. D. Palmer, "Human Rhythms," *Bioscience* 27 (February 1977): 93-99.
23. L. E. Scheving, "Mitotic Activity in Human Epidermis," *Anatomical Record* 135 (1959): 7–20.
24. Palmer, *Introduction to Biological Rhythms,* p. 122.
25. Jürgen Aschoff, Klauss Hoffmann, Hermann Pohl, and Rütger Wever, "Re-entrainment of Circadian Rhythms after Phase Shift of the Zeitgeber," *Chronobiologia* 2 (1975): 23–78.
26. Luce, *Biological Rhythms,* p. 122.
27. Ibid., p. 129.
28. M. H. Smolensky, P. Hartsock, M. C. Lu, S. Stevens, M. Lagoguey, and A. Reinberg, "Human Circannual Rhythms of Birth and Reproductive Function," unpublished paper.
29. Michael Smolensky, Franz Halberg, and Frederick Sargent II, "Chronobiology of the Life Sequence," in *Advances in Climatic Physiology,* ed. S. Ito, K. Ogata, and H. Yoshimura (Tokyo: Igaku Shoin Ltd., 1972).
30. Ibid.
31. Ibid.
32. Smolensky, Hartsock, Lu, Stevens, Lagoguey, and Reinberg, "Human Circannual Rhythms."
33. Ibid.
34. Luce, *Biological Rhythms.*
35. Smolensky, Hartsock, Lu, Stevens, Lagoguey, and Reinberg, "Human Circannual Rhythms."
36. Mills, ed., *Biological Aspects of Circadian Rhythms.*
37. Smolensky, Hartsock, Lu, Stevens, Lagoguey, and Reinberg, "Human Circannual Rhythms."
38. Palmer, "Human Rhythms."
39. Ibid.
40. Luce, *Biological Rhythms.*
41. Milton H. Stetson and Marcie Watson-Whitmyre, "Nucleus Suprachiasmaticus: The Biological Clock in the Hamster?" *Science,* January 16, 1976.
42. Martin C. Moore-Ede and Frank M. Sulzman, "The Physiological Basis of Circadian Timekeeping in Primates," remarks at the Federation Meetings, 1977.
43. Palmer, *Introduction to Biological Rhythms.*

44. Martin C. Moore-Ede and Frank M. Sulzman, "Internal Temporal Order," in *Handbook of Behavioral Neurobiology,* vol. 5 ed. Jürgen Aschoff (New York: Plenum Press, forthcoming).

Chapter Four

1. Peter Hauri, "The Sleep Disorders," Current Concepts series (Kalamazoo, Mich.: The Upjohn Company, 1977), p. 9.
2. Dement, *Some Must Watch,* p. 25.
3. Hauri, "The Sleep Disorders," pp. 6–8.
4. Dement, *Some Must Watch,* p. 79.
5. Hauri, "The Sleep Disorders," p. 65.
6. Dement, *Some Must Watch,* p. 79.
7. Luce, *Biological Rhythms,* p. 18.
8. Dement, *Some Must Watch,* pp. 77–78.
9. Ennio Vivaldi, J. Allan Hobson, and Robert L. McCarley, press release from Association for the Psychophysiological Study of Sleep, Stanford University, April 1978.
10. Dement, *Some Must Watch,* p. 50.
11. Ibid., p. 71.
12. Hauri, "The Sleep Disorders," p. 15.
13. Wilse B. Webb, remarks at Association for the Psychophysiological Study of Sleep, Stanford University, April 1978.
14. Dement, *Some Must Watch,* p. 5.
15. Webb, remarks at Association for Psychophysiological Study of Sleep.
16. Hauri, "The Sleep Disorders," p. 14.
17. Ibid., p. 17.
18. Conversations with staff of Montefiore Hospital, New York, February–November 1978.
19. Hauri, "The Sleep Disorders," p. 19.
20. Conversation with Paul Naitoh of the Naval Health Research Center, San Diego, Calif., July 1978.
21. Dement, *Some Must Watch,* p. 6.
22. Conroy and Mills, *Human Circadian Rhythms,* pp. 94–95.
23. Dement, *Some Must Watch,* p. 7.
24. Ibid., p. 8.
25. Ibid.

26. Ibid., p. 13.
27. Ibid., pp. 8 ff., 87.
28. Hauri, "The Sleep Disorders," pp. 25–27.
29. Ibid.
30. Ibid., pp. 22–27.
31. Ibid.
32. Ibid., pp. 53–54.
33. Luce, *Biological Rhythms,* pp. 28–29.
34. Peter J. Hauri and Peter M. Silverfarb, press release of the Association for the Psychophysiological Study of Sleep, Stanford University, April 1978.
35. Hauri, "The Sleep Disorders," p. 5.
36. Luce, *Biological Rhythms,* p. 18.
37. Hauri, "The Sleep Disorders," p. 10.
38. Ibid., p. 61.
39. Ibid., p. 24.
40. Ibid., p. 32.
41. Ibid., p. 29.
42. Dement, *Some Must Watch,* p. 80.
43. Ibid., p. 81.
44. Ibid., p. 92.
45. Luce, *Biological Rhythms,* p. 23.
46. Bruce J. Durie, remarks at Association for the Psychophysiological Study of Sleep, Stanford University, April 1978.

Chapter Five

1. Luce, *Biological Rhythms,* p. 96.
2. Richard J. Wurtman and Michael A. Moskowitz, "The Pineal Organ," *New England Journal of Medicine* 296 (June 9, 16, 1977).
3. V. R. Mountcastle, ed., *Medical Physiology,* vol. 2 (St. Louis: C. V. Mosby Company, 1974), pp. 1697–98.
4. Laurence E. Scheving, Heinz v. Mayersbach, and John E. Pauly, "An Overview of Chronopharmacology," *Journal Européen de Toxicologie* (July–August 1974).

5. Ronald Hurst, ed., *Pilot Error* (London: Crosby Lockwood Staples, 1976), p. 83.
6. Luce, *Biological Rhythms*, p. 97.
7. Ibid., p. 90.
8. Ibid., p. 26.
9. Walter McQuade and Ann Aikman, *Stress* (New York: E. P. Dutton, 1974), p. 4.
10. Ibid., p. 6.
11. Charles Swencionis, Langley-Porter Neuro-Psychiatric Institute, University of California, San Francisco, remarks at "Concepts of Stress" Symposium, Los Angeles, February 25–26, 1978.
12. Ibid.
13. Joan Vernikos-Danellis, NASA/Ames Research Center, Moffett Field, Calif., interview, March 1978.
14. Ibid.
15. Luce, *Biological Rhythms*, pp. 116–117.
16. Peter Blythe, *Stress Disease* (New York: St. Martin's Press, 1973).
17. Swencionis, remarks at "Concepts of Stress" Symposium, Los Angeles, February 25–26, 1978.
18. Ray Rosenman, remarks at "Concepts of Stress" Symposium, Los Angeles, February 25–26, 1978.
19. Ibid.
20. Ibid.
21. Ibid.
22. Ibid.
23. Ibid.
24. Conversation with Flight Surgeon Pattee, U.S. Naval Air Station, El Toro, Calif.
25. Rosenman, remarks at "Concepts of Stress" Symposium, Los Angeles, February 25–26, 1978.
26. Ibid.
27. Ibid.
28. Michael Gideon Marmot, "Acculturation and Coronary Heart Disease in Japanese-Americans," abstract (Berkeley: University of California, School of Public Health, 1978); and S. Leonard Syme, Warren Winkelstein, Jr., and Michael G. Marmot, press release, University of California, Berkeley.

29. Thomas H. Holmes, remarks at "Concepts of Stress" Symposium, Los Angeles, February 25–26, 1978.
30. Cheryl Hart, University of Washington, interview, August 1978.

Chapter Six

1. Palmer, *Introduction to Biological Rhythms,* p. 139.
2. John R. Beljan, *Human Performance in the Aviation Environment,* Part I-A, NAS2–6657 (Washington D.C.: National Aeronautics and Space Administration, 1972), pp. 209–210.
3. Swencionis, remarks at "Concepts of Stress" Symposium, Los Angeles, February 25–26, 1978.
4. Frederick Hegge, Walter Reed Hospital, interview, August 1978.
5. Charles M. Winget, Lewis Hughes, and Joseph LaDou, "Physiological Effects of Rotational Work Shifting: A Literature Survey," *Journal of Occupational Medicine* 20, no. 3 (March 1978).
6. Donald L. Tasto, Michael J. Colligan, Eric W. Skjei, and Susan J. Polly, *Health Consequences of Shift Work* (Cincinnati: National Institute for Occupational Safety and Health, Robert A. Taft Laboratories, 1978).
7. Ibid.
8. Aschoff, Hoffmann, Pohl, and Wever, "Re-entrainment of Circadian Rhythms."
9. Conroy and Mills, *Human Circadian Rhythms,* p. 152.
10. Elliot D. Weitzman, "Effect of Sleep-Wake Cycle Shifts on Sleep and Neuroendocrine Function," in *Behavior and Brain Electrical Activity,* ed. N. Burch and H. L. Altshuler (New York: Plenum Publishing Corporation).
11. Winget, Hughes, and LaDou, "Physiological Effects of Rotational Work Shifting."
12. Ibid.
13. Conroy and Mills, *Human Circadian Rhythms,* pp. 104, 150.
14. Elliot D. Weitzman, Montefiore Hospital, Albert Einstein College of Medicine, interview, February 1978.
15. Tasto, Colligan, Skjei, and Polly, *Health Consequences of Shift Work.*

16. Conroy and Mills, *Human Circadian Rhythms,* pp. 149-150.
17. Winget, Hughes, and LaDou, "Physiological Effects of Rotational Work Shifting."
18. Ibid.
19. Stanley D. Nollen and Virginia H. Martin, *Alternative Work Schedules, Part 1: Flexitime* (New York: American Management Associations, 1978).
20. Ibid.
21. Raloff, "Biological Clocks," p. 62.
22. Frederick G. Hofmann and Adele D. Hofmann, *A Handbook on Drug and Alcohol Abuse* (New York: Oxford University Press, 1975), p. 102.
23. *Alcohol and Health,* First Special Report to the U.S. Congress from the Secretary of Health, Education and Welfare, December 1971, p. 6.
24. Luce, *Biological Rhythms,* p. 75.
25. Palmer, "Human Rhythms."
26. Emanuel Rubin, Charles S. Lieber, Kurt Altman, Gary G. Gordon, and A. Louis Southern, "Prolonged Alcohol Consumption Increases Testosterone Metabolism in the Liver," *Science,* February 13, 1976.
27. Conroy and Mills, *Human Circadian Rhythms.*
28. Vernikos-Danellis, interview, March 1978.
29. F. Halberg, Erna Halberg, and Franca Carandente, "Chronobiology and Metabolism in the Broader Context of Timely Intervention and Timed Treatment," in *Diabetes Research Today,* Symposia Medica Hoechst 12, Professor Dr. A. Renold, Prof. Dr. W. Creutzfeldt, and Prof. Dr. E. F. Pfeiffer, Chairmen.
30. Michael H. Smolensky, Sheryl E. Tatar, Stuart A. Bergman, Jacques G. Losman, Christian N. Barnard, Clifford C. Dacso, and Irving A. Kraft, "Circadian Rhythmic Aspects of Human Cardiovascular Function: A Review by Chronobiologic Statistical Methods," *Chronobiologia* 3 (October–December 1976).

Chapter Seven

1. Franz Halberg, "Implications of Biologic Rhythms for Clinical Practice," *Hospital Practice,* January 1977, pp. 139–149.

2. Luce, *Biological Rhythms.*
3. Ibid.
4. Martin C. Moore-Ede, "Circadian Rhythms of Drug Effectiveness and Toxicity," *Clinical Pharmacology and Therapeutics,* November-December 1973, p. 930.
5. Luce, *Biological Rhythms.*
6. Ibid., p. 78.
7. Ibid., p. 75.
8. L. E. Scheving, D. F. Bedral, and J. E. Pauly, "Daily Circadian Rhythm in Rats to D-Amphetamine Sulfate: Effect of Blinding and Continuous Illumination of the Rhythm," *Nature* 219 (1968): 621–22.
9. Smolensky, Tatar, Bergman, Losman, Barnard, Dacso, and Kraft, "Circadian Rhythmic Aspects of Human Cardiovascular Function."
10. H. v. Mayersbach, "Time—A Key in Experimental and Practical Medicine," *Archives of Toxicology* 36 (1976).
11. *New York Times,* December 5, 1978, p. C1.
12. Daniel F. Kripke, Daniel J. Mullaney, Martha Atkinson, and Sanford Wolf, "Circadian Rhythm Disorders in Manic-Depressives," *Biological Psychiatry* 13, no. 3 (1978).
13. Luce, *Biological Rhythms,* p. 49.
14. Ibid., p. 18.
15. D. A. Rockwell, C. M. Winget, L. S. Rosenblatt, E. A. Higgins, and N. W. Hetherington, "Biological Aspects of Suicide—Circadian Disorganization," *Journal of Nervous Mental Diseases* 166, no. 12 (1978).
16. Luce, *Biological Rhythms,* p. 112.
17. Ibid., pp. 5-6.
18. Dement, *Some Must Watch.*
19. Hauri, "The Sleep Disorders."
20. Ibid., p. 46.
21. Dement, *Some Must Watch.*
22. Hauri, "The Sleep Disorders," p. 47.
23. Ismet Karacan, F. Brantley Scott, Patricia J. Salis, Samuel L. Attia, J. Catesby Ware, Attila Altinel, and Robert L. Williams, "Nocturnal Erections, Differential Diagnosis of Impotence, and Diabetes," *Biological Psychiatry* 12, no. 3 (1977): 374.

24. Charles F. Ehret and Kenneth W. Dobra, "The Oncogenic Implications of Chronobiotics in the Synchronization of Mammalian Circadian Rhythms: Barbiturates and Methylated Xanthines," *Proceedings of the Third Annual International Symposium on the Detection and Prevention of Cancer,* ed. H. E. Nieburgs (New York: Marcel Dekker, 1977), pp. 1101–1114.
25. H. W. Simpson, Royal Infirmary, Glasgow, Scotland, interview, September 1978.
26. L. E. Scheving, E. R. Burns, and J. E. Pauly, "Can Chronobiology Be Ignored When Considering the Cancer Problem?" in *Prevention and Detection of Cancer,* ed. H. E. Nieburgs (New York: Marcel Dekker, 1977), pp. 1063–1079.
27. Halberg, "Implications of Biologic Rhythms for Clinical Practice."

Chapter Eight

1. Beljan, *Human Performance in the Aviation Environment,* p. 210.
2. Charles Winget, NASA/Ames Research Center, interview, June 1977.
3. Hubertus Strughold, *Your Body Clock* (New York: Charles Scribner's Sons, 1971), p. 48.
4. Dement, *Some Must Watch,* p. 6.
5. Winget, interview, June 1977.
6. Strughold, *Your Body Clock.*
7. Beljan, *Human Performance in the Aviation Environment.*
8. Workshop on "Biological Clocks," Stanford University.
9. Stanley R. Mohler, J. Robert Dille, Harry L. Gibbons, "Circadian Rhythms and the Effects of Long Distance Flying," Office of Aviation Medicine, Federal Aviation Administration, April 1968.
10. Ross A. McFarland, "Influence of Changing Time Zones on Air Crews and Passengers," *Aerospace Medicine,* June 1974.
11. Charles F. Ehret, Kenneth R. Groh, and John C. Meinert, "Circadian Dysynchronism and Chronotypic Ecophilia as Factors in Aging and Longevity," *Advances in Experimental Medicine and Biology* (New York: Plenum Press).

12. Hauri, "The Sleep Disorders."
13. Aschoff, Hoffmann, Pohl, and Wever, "Human Circadian Rhythms."
14. Beljan, *Human Performance in the Aviation Environment*, p. 209.
15. Workshop on "Biological Clocks," Stanford University.
16. Luce, *Biological Rhythms*, p. 137.

Chapter Nine

1. *Los Angeles Times*, May 24, 1977.
2. K. E. Klein, H. M. Wegmann, G. Athanassenas, H. Hohlweck, and P. Kuklinski, "Air Operations and Circadian Performance Rhythms," *Aviation Space Environmental Medicine* 47 (1976): 221-230.
3. Hurst, ed., *Pilot Error*.
4. Ibid., p. 13.
5. Ministry of Transportation, Communications and Civil Aeronautics, Republic of Bolivia, Report of Boeing 707 accident, October 13, 1976 (report date: January 31, 1977), Santa Cruz de la Sierra, Bolivia.
6. Ibid.
7. Hurst, ed., *Pilot Error*, p. 27.
8. Ibid., p. 20.

Bibliography

Adamich, Maria; Laris, Philip C.; and Sweeney, Beatrice M. "Switched-on Membranes: Internal Clocks." *Science News*, July 10, 1976.

Alcohol and Health. First Special Report to the U.S. Congress from the Secretary of Health, Education and Welfare, December 1971.

Apfelbaum, M.; Reinberg, A.; and Lucatis, D. "Meal Timing and Human Circadian Rhythms." *International Journal of Chronobiology* 4 (1976): 29–37.

Aschoff, Jürgen. "Circadian Rhythms in Man." *Science* 148 (1965): 1427–1432.

Aschoff, Jürgen. "Desynchronization and Resynchronization of Human Circadian Rhythms." *Aerospace Medicine* 40 (1969): 844–849.

Aschoff, J.; Fatranska, M.; Giedke, H.; Doerr, P.; Stamm, D.; and Wisser, H. "Human Circadian Rhythms in Continuous Darkness: Entrainment by Social Cues." *Science* 171 (1971).

Aschoff, Jürgen; Hoffmann, Klauss; Pohl, Hermann; and Wever, Rütger. "Re-entrainment of Circadian Rhythms after Phase Shift of the Zeitgeber." *Chronobiologia* 2 (1975): 23–78.

Aschoff, J., and Pohl, H. "Rhythmic Variations in Energy Metabolism." *Federation Proceedings* 29 (1970): 1541–1552.

Aschoff, Jürgen, and Wever, Rütger. "Human Circadian Rhythms: A Multioscillator System." *Federation Proceedings* 35 (1976): 2326–2332.

Bafitis, Harold; Smolensky, Michael H.; Hsi, Bartholomew P.; Mahoney, Steven; Schectman, Tommy; Kresse, Herman; Powell, Susan; and Dutton, Laverne. "A Circadian Susceptibility/Resistance Rhythm for Potassium Cyanide in Male Balb/cCr Mice." Unpublished paper.

Beljan, John R. *Human Performance in the Aviation Environment,* Part I-A, NAS2–6657. Washington, D.C.: National Aeronautics and Space Administration, 1972.

Binkley, S.; Riebman, J. B.; and Reilly, K. B. "Timekeeping by the Pineal Gland." *Science,* September 16, 1977, pp. 1181–1183.

Blythe, Peter. *Stress Disease.* New York: St. Martin's Press, 1973.

Březinova, Vlasta. "Effect of Caffeine on Sleep: EEG Study in Late Middle Age People." *British Journal of Clinical Pharmacology* 1 (1974): 203–208.

Brown, Frank A., Jr. "The 'Clocks' Timing Biological Rhythms." *American Scientist* 60 (November-December 1972): 756–766.

Burns, E. Robert, and Scheving, Laurence E. "The Influence of Circadian Variation on *In Vivo* Cell Kinetic Studies." In *Prevention and Detection of Cancer,* edited by H. E. Nieburgs, New York: Marcel Dekker, 1978.

Cohen, E. L., and Wurtman, R. J. "Brain Acetylcholine." *Science,* February 13, 1976.

Conroy, R. T. W. L., and Mills, J. N. *Human Circadian Rhythms.* London: J. & A. Churchill, 1970.

Crowley, Thomas J.; Lubanovic, William; Halberg, Franz; and Hydinger, Marilyn. "Daily Oral Methadone Alters Circadian Activity Rhythm of Monkeys." *Proceedings of the 12th International Conference,* International Society of Chronobiology, Washington, D.C. Milan, Italy: Il Ponte, 1977.

Dement, William C. *Some Must Watch While Some Must Sleep.* San Francisco: San Francisco Book Company, Inc., 1976.

Dement, William C., and Baird, William P. *Narcolepsy: Care and Treatment.* Stanford, Calif.: The American Narcolepsy Association, 1977.

Descovich, Gian Carlo; Montalbetti, Norberto; Frederich, Jurgen; Kuhn, Wilhelm; Rimondi, Silvana; Halberg, Franz; and Ceredi,

Carla. "Age and Catecholamine Rhythms." *Chronobiologia* 1, no. 2 (April-June 1974).

DeWied, David. "Pituitary-Adrenal System Hormones and Behavior." In *The Neurosciences*, edited by Francis O. Schmitt and Frederick G. Worden. Cambridge, Mass.: The MIT Press, 1974.

Dexter, James D., and Weitzman, Elliot D. "The Relationship of Nocturnal Headaches to Sleep Stage Patterns." *Neurology* 20, no. 5 (May 1970).

Ehret, Charles F. "The Sense of Time: Evidence for Its Molecular Basis in the Eukaryotic Gene-Action System." *Advances in Biological and Medical Physics*, vol. 15. New York: Academic Press, Inc., 1974.

Ehret, Charles F., and Dobra, Kenneth W. "The Oncogenic Implications of Chronobiotics in the Synchronization of Mammalian Circadian Rhythms: Barbiturates and Methylated Xanthines." In *Proceedings of the Third International Symposium on the Detection and Prevention of Cancer*, edited by H. E. Nieburgs. New York: Marcel Dekker, 1977.

Ehret, Charles F.; Groh, Kenneth R.; and Meinert, John C. "Circadian Dysynchronism and Chronotypic Ecophilia as Factors in Aging and Longevity." In *Advances in Experimental Medicine and Biology*. New York: Plenum Press, forthcoming.

Ehret, C. F.; Potter, V. R.; and Dobra, K. W. "Chronotypic Action of Theophylline and of Pentobarbital as Circadian Zeitgebers in the Rat." *Science*, June 10, 1975, pp. 1212–1215.

Elliott, Jeffrey A. "Circadian Rhythms and Photoperiodic Time Measurement in Mammals." *Federation Proceedings* 35 (1976): 2339–2346.

Feldman, Jerry F. "Circadian Periodicity in Neurospora." *Science*, November 25, 1975.

Fernstrom, John D. "Effects of the Diet on Brain Neurotransmitters." *Metabolism* 26, no. 2 (February 1977).

Fernstrom, John D., and Wurtman, Richard J. "Nutrition and the Brain." *Scientific American*, February 1974.

Frohberg, H. "Introduction to the Symposium 'What Is the Importance of Chronobiology in Toxicology and Pharmacology?'" *Archives of Toxicology* 36, nos. 3 and 4 (1976).

Fuller, Charles A.; Sulzman, Frank M.; and Moore-Ede, Martin C. "Thermoregulation Is Impaired in an Environment Without

Circadian Time Cues." *Science* 199 (February 17, 1978): 794–796.

Gautherie, Michel, and Gros, Charles. "Circadian Rhythm Alteration of Skin Temperature in Breast Cancer." *Chronobiologia* 4 (January–March 1977).

Gould, Stephen Jay. "The 120-Year Bamboo Clock." *Natural History,* April 1, 1977.

Halberg, Franz. "Implications of Biologic Rhythms for Clinical Practice." *Hospital Practice,* January 1977.

Halberg, Franz; Carandente, Franca; Cornelissen, Germaine; and Katinas, George S. *Glossary of Chronobiology.* Little Rock, Ark.: Department of Anatomy, University of Arkansas Medical Center, 1977.

Halberg, F.; Gupta, B. D.; Haus, E.; Halberg, E.; Deka, A. C.; Nelson, W.; Sothern, R. B.; Kornelissen, G.; Klee, J.; Lakatua, D. J.; Scheving, L. E.; and Burns, E. R. "Steps Toward a Cancer Chronotherapy." *XIVth International Congress of Therapeutics.* Montpellier, France: L'Expansion Scientifique Française, 1977.

Halberg, F.; Halberg, Erna; and Carandente, Franca. "Chronobiology and Metabolism in the Broader Context of Timely Intervention and Timed Treatment." In *Diabetes Research Today,* Symposia Medica Hoechst 12, edited by Prof. Dr. A. Renold, Prof. Dr. W. Creutzfeldt, and Prof. Dr. E. F. Pfeiffer, Stuttgart–New York: F. K. Schattauer Verlag, 1976.

Halberg, Franz; Nelson, Walter L.; and Cadotte, Linda. "Living Routine Shifts Simulated on Mice by Weekly or Twice-Weekly Manipulation of Light-Dark Cycle." *Proceedings of the 12th International Conference,* International Society of Chronobiology, Washington, D.C. Milan, Italy: Il Ponte, 1977.

Hale, Henry B.; Hartman, Bryce O.; Harris, Dickie A.; Miranda, Roberto E.; and Williams, Edgar W. "Physiologic Cost of Prolonged Double-Crew Flights in C-5 Aircraft." *Aerospace Medicine,* September 1973.

Harper, C. R., and Kidera, G. J. "Aviator Performance and the Use of Hypnotic Drugs." *Aerospace Medicine,* February 1972.

Hartman, Bryce O.; Hale, Henry B.; Harris, Dickie A.; and Sanford, James F., III. "Psychobiologic Aspects of Double-Crew Long-

Duration Missions in C-5 Aircraft." *Aerospace Medicine,* October 1974.

Hauri, Peter. "The Sleep Disorders." Current Concepts series. Kalamazoo, Mich.: The Upjohn Company, 1977.

Hofmann, Frederick G., and Hofmann, Adele D. *A Handbook on Drug and Alcohol Abuse.* New York: Oxford University Press, 1975.

Hurst, Ronald, ed. *Pilot Error.* London: Crosby Lockwood Staples, 1976.

Jovet, Michel; Mouret, Jacques; Chouvet, Guy; and Siffre, Michel. "Toward a 48-Hour Day." In *The Neurosciences,* edited by Francis O. Schmitt and Frederick G. Worden. Cambridge, Mass.: The MIT Press, 1974.

Kanabrocki, Col. E. L.; Scheving, Col. L. E.; Halberg, Prof. F.; Brewer, Lt. Col. R. L.; and Bird, Lt. Col. T. J. "Circadian Variation in Presumably Healthy Young Soldiers." Springfield, Va.: National Technical Information Service, U.S. Department of Commerce, 1974.

Karacan, Ismet. "Advances in the Psychophysiological Evaluation of Male Erectile Impotence." Weekly Psychiatry Update Series, no. 43. New York: Biomedia, Inc., 1977.

Karacan, Ismet; Anch, A. Michael; and Williams, Robert L. "Recent Advances in the Psychophysiology of Sleep and Their Psychiatric Significance." In *Biological Foundations of Psychiatry,* edited by R. G. Grenell and S. Gabay. New York: Raven Press, 1976.

Karacan, Ismet; Salis, Patricia; Thornby, John I.; and Williams, Robert L. "The Ontogeny of Nocturnal Penile Tumescence." *Waking and Sleeping* 1 (1976): 27–44.

Karacan, Ismet; Scott, F. Brantley; Salis, Patricia J.; Attia, Samuel L.; Ware, J. Catesby; Altinel, Attila; and Williams, Robert L. "Nocturnal Erections, Differential Diagnosis of Impotence, and Diabetes." *Biological Psychiatry* 12, no. 3 (1977).

Kavanau, J. Lee, and Peters, Charles R. "Activity of Nocturnal Primates: Influences of Twilight Zeitgebers and Weather." *Science,* January 9, 1976.

Klein, K. E.; Brüner, H.; Holtmann, H.; Rehme, H.; Stolze, J.; Steinhoff, W. D.; and Wegman, H. M. "Circadian Rhythm of

Pilots' Efficiency and Effects of Multiple Time Zone Travel." *Aerospace Medicine,* February 1970.

Klein, K. E.; Wegmann, H. M.; Athanassenas, G.; Hohlweck, H.; and Kuklinski, P. "Air Operations and Circadian Performance Rhythms." *Aviation, Space, Environmental Medicine* 47 (1976): 221-230.

Klein, K. E.; Wegmann, H. M.; and Hunt, Bonnie I. "Desynchronization of Body Temperature and Performance Circadian Rhythm as a Result of Outgoing and Homegoing Transmeridian Flights." *Aerospace Medicine,* February 1972.

Kositskiy, G. I., and Smirnov, V. M. *The Nervous System and "Stress."* Moscow: "Nanka" Press, 1970. Translated from the Russian and published by NASA, 1972.

Kramer, Milton. "Manifest Dream Content in Normal and Psychopathologic States." *Archives of General Psychiatry,* February 1970.

Kramer, Milton; Roehrs, Timothy; and Roth, Thomas. "Mood Change and the Physiology of Sleep." *Comprehensive Psychiatry,* January-February 1976.

Kramer, M., and Roth, T. "The Mood-Regulating Function of Sleep." In *Sleep,* edited by W. P. Koella and P. Levin. Basel: S. Karger AG, 1973.

Kramer, Milton; Roth, Thomas; and Cisco, John. "The Meaningfulness of Dreams." In *Sleep 1976,* edited by W. P. Koella and P. Levin. Basel: S. Karger AG, 1977.

Kramer, Milton; Roth, Thomas; and Czaya, John. "Dream Development Within a REM Period." In *Sleep 1974,* edited by P. Levin and W. P. Koella. Basel: S. Karger AG, 1975.

Kripke, Daniel F.; Mullaney, Daniel J.; Atkinson, Martha; and Wolf, Sanford. "Circadian Rhythm Disorders in Manic-Depressives." *Biological Psychiatry* 13, no. 3 (1978).

Land, J. W. "Role of Circadian Rhythms in Alligators." *Science,* February 13, 1976.

Lang, Jefferey W. "Amphibious Behavior of *Alligator Mississippiensus:* Role of a Circadian Rhythm and Light." *Science,* February 13, 1976.

Levett, Michael A. "Pilot Error—Studies Seek Out Causes." *Los Angeles Times,* May 24, 1977.

Levine, Howard; Halberg, Erna; Halberg, Franz; Thompson, Mark E.; Graeber, R. Curtis; Thompson, Dean; and Jacobs, Harry L. "Changes in Internal Timing of Heart Rate, Diastolic Blood Pressure and Certain Aspects of Physical and Mental Performance in Presumably Healthy Subjects on Different Meal Schedules." *Proceedings of the 12th International Conference,* International Society of Chronobiology, Washington, D.C. Milan, Italy: Il Ponte, 1977.

Luce, Gay Gaer. *Biological Rhythms in Psychiatry and Medicine.* Washington, D.C.: National Institute of Mental Health, U.S. Department of Health, Education and Welfare, 1970.

Lynch, H. J.; Ozaki, Y.; Shakal, D.; and Wurtman, R. J. "Melatonin Excretion of Man and Rats: Effect of Time of Day, Sleep, Pinealectomy and Food Consumption." *International Journal of Biometeorology* 19, no. 4 (1975).

Lynch, H. J.; Wurtman, R. J.; Moskowitz, M. A.; Archer, M. C.; and Ho, M. H. "Daily Rhythms in Human Urinary Melatonin." *Science,* January 17, 1975.

McFarland, Ross A. "Air Travel across Time Zones." *American Scientist,* January-February 1975.

McFarland, Ross A. "Influence of Changing Time Zones on Air Crews and Passengers." *Aerospace Medicine,* June 1974, pp. 648–658.

Malewiak, M. Irene; Griglio, Sabin; Halberg, Franz; Nelson, Walter L.; Apfelbaum, Marian; and Kalopissis, Anna D. "Circadian Temperature Rhythm and Biochemical Findings in Rats on Different Diets with or Without Carbohydrate and with Inversely Varying Contents of Protein and Fat." *Proceedings of the 12th International Conference,* International Society of Chronobiology, Washington, D.C. Milan, Italy: Il Ponte, 1977.

Marmot, Michael Gideon. "Acculturation and Coronary Heart Disease in Japanese-Americans." Abstract. Berkeley: University of California, School of Public Health, 1978.

Mayer, Dieter. "The Circadian Rhythm of Synthesis and Catabolism of Cholesterol." *Archives of Toxicology* 36, nos. 3 and 4 (1976).

Mayersbach, H. v. "Time—A Key in Experimental and Practical Medicine." *Archives of Toxicology* 36 (1976).

Meinert, John C.; Ehret, Charles F.; and Antipa, Gregory A. "Circadian Chronotypic Death in Heat-Synchronized Infradian Mode Cultures of *Tetrahymena pyriformis W.*" *Microbial Ecology* 2 (1975): 201–214.

Menaker, Michael. "Aspects of the Physiology of Circadian Rhythmicity in the Vertebrate Central Nervous System." In *The Neurosciences,* edited by Francis O. Schmitt and Frederick G. Worden. Cambridge, Mass.: The MIT Press, 1974.

Miles, L. E. M.; Raynal, D. M., and Wilson, M. A. "Blind Man Living in Normal Society Has Circadian Rhythms of 24.9 Hours." *Science,* October 28, 1977.

Mills, J. N., ed. *Biological Aspects of Circadian Rhythms.* London and New York: Plenum Press, 1973.

Mohler, Stanley R. "Physiological Index as an Aid in Developing Airline Pilot Scheduling Patterns." *Aviation, Space, Environmental Medicine,* March 1976.

Mohler, Stanley R.; Dille, J. Robert; and Gibbons, Henry L. "The Time Zone and Circadian Rhythms in Relation to Aircraft Occupants Taking Long-Distance Flights." *American Journal of Public Health,* August 1968.

Moore, Robert Y. "Visual Pathways and the Central Neural Control of the Diurnal Rhythms." In *The Neurosciences,* edited by Francis O. Schmitt and Frederick G. Worden. Cambridge, Mass.: The MIT Press, 1974.

Moore-Ede, Martin C. "Circadian Rhythms of Drug Effectiveness and Toxicity." *Clinical Pharmacology and Therapeutics,* November–December 1973.

Moore-Ede, Martin C. "The Physiological Basis of Circadian Timekeeping in Primates." Remarks at Federation Meetings, 1977.

Moore-Ede, Martin C., and Sulzman, Frank M. "Internal Temporal Order." In *Handbook of Behavioral Neurobiology,* vol. 5: *Biological Rhythms,* edited by Jürgen Aschoff, New York: Plenum Press, forthcoming.

Morin, Lawrence P.; Fitzgerald, Kathleen M.; and Zucker, Irving. "Estradiol Shortens the Period of Hamster Circadian Rhythms." *Science,* April 15, 1977.

Mountcastle, V. R., ed. *Medical Physiology,* vol. 2. St. Louis: The C. V. Mosby Company, 1974.

Nathanson, James A., and Greengard, Paul. "'Second Messengers' in the Brain." *Scientific American,* July 29, 1977.

Netter, Frank. *The Endocrine System.* The CIBA Collection, 1965.
"New Facts on Biorhythms." *Science Digest,* May 1976.
Nollen, Stanley D., and Martin, Virginia H. *Alternative Work Schedules, Part 1: Flexitime.* New York: American Management Associations, 1978.
Orford, Robert R., and Carter, Earl T. "Survival as an Airline Pilot." San Francisco: Aerospace Medical Association, May 1, 1975.
Palmer, John D. "Biological Clocks of the Tidal Zone." *Scientific American,* February 1975, pp. 70–77.
Palmer, J. D. "Human Rhythms." *Bioscience* 27 (February 1977): 93–99.
Palmer, John D. *Introduction to Biological Rhythms.* New York: Academic Press, Inc., 1976.
Pittendrigh, Colin S. "Circadian Oscillations in Cells and the Circadian Organization of Multicellular Systems." In *The Neurosciences,* edited by Francis O. Schmitt and Frederick G. Worden. Cambridge, Mass.: The MIT Press, 1974.
Raloff, Janet. "Biological Clocks: How They Affect Your Health." *Scientific Digest,* November 1975, pp. 62–69.
Reinberg, Alain; Halberg, Franz; and Falliers, Constantine J. "Circadian Timing of Methylprednisolone Effects in Asthmatic Boys." *Chronobiologia* 1, no. 4 (October–December 1974).
Reinberg, Alain; Zagula-Mally, Zenona; Ghata, Jean; and Halberg, Jean. "Circadian Reactivity Rhythm of Human Skin to House Dust, Penicillin, and Histamine." *The Journal of Allergy,* November 1969.
Rockwell, Don A.; Hodgson, Michael G.; Beljan, John R.; and Winget, Charles M. "Psychologic and Psychophysiologic Response to 105 Days of Social Isolation." *Aviation, Space, and Environmental Medicine,* October 1976.
Rockwell, D. A.; Winget, C. M.; Rosenblatt, L. S.; Higgins, E. A.; and Hetherington, N. W. "Biological Aspects of Suicide—Circadian Disorganization." *Journal of Nervous and Mental Disease* 166, no. 12 (1978).
Roscoe, Alan H. "Stress and Workload in Pilots." *Aviation, Space, and Environmental Medicine,* April 1978.
Roth, Thomas; Kramer, Milton; and Roehrs, Timothy. "Mood Before and After Sleep." *The Psychiatric Journal of the University of Ottawa,* November 1976.

Rubin, Emanuel; Lieber, Charles S.; Altman, Kurt; Gordon, Gary G.; and Southern, A. Louis. "Prolonged Alcohol Consumption Increases Testosterone Metabolism in the Liver." *Science,* February 13, 1976.

Rusak, Benjamin, and Zucker, Irving. "Biological Rhythms and Animal Behavior." In *Annual Review of Psychology.* Palo Alto, Calif.: Annual Reviews, Inc., 1975.

Santandreu, Herrero Aldama. "Casualties Due to Sickness and Incapacitation in Aviation." Iberia Lineas Areas de España. Translated by Paul Keller. Unpublished paper.

Saunders, D. S. "The Biological Clock of Insects." *Scientific American,* February 1976.

Saunders, David S. *An Introduction to Biological Rhythms.* New York: John Wiley and Sons, 1977.

Scheving, Laurence E. "The Dimension of Time in Biology and Medicine—Chronobiology." *Endeavour,* May 1976.

Scheving, L. E.; Burns, E. R.; and Pauly, J. E. "Can Chronobiology Be Ignored When Considering the Cancer Problem?" In *Prevention and Detection of Cancer,* edited by H. E. Nieburgs. New York: Marcel Dekker, 1977.

Scheving, Laurence E.; Burns, E. Robert; Pauly, John E.; Halberg, Franz; and Haus, Erhard. "Survival Cure of Leukemic Mice after Circadian Optimization of Treatment with Cyclophosphamide and 1-β-D-Arabinofuranosylcytosine." *Cancer Research,* October 1977.

Scheving, Laurence E.; Burns, E. Robert; Pauly, John E.; and Tsai, Tien-Hu. "Circadian Variation in Cell Division of the Mouse Alimentary Tract, Bone Marrow and Corneal Epithelium." *The Anatomical Record* 191 (August 1978).

Scheving, Laurence E.; Mayersbach, Heinz v.; and Pauly, John E. "An Overview of Chronopharmacology." *Journal Européen de Toxicologie,* July-August 1974.

Scheving, Laurence E., and Pauly, John E. "Circadian Rhythms: Some Examples and Comments on Clinical Application." *Chronobiologia* 1 (January-March 1974).

Schmeck, Harold M., Jr. "Manic-Depressive Cycle Tied to 'Clock' Defect." *New York Times,* December 5, 1978.

Schmidt-Koenig, Klaus. *Migration and Homing in Animals.* Berlin–Heidelberg–New York: Springer-Verlag, 1975.

Sekiguchi, Chiharu; Yamaguchi, Ototsugu; Kitajima, Takeyuki; and Ueda, Yasushi. "The Effects of Rapid Round Trips Against Time Displacement on Adrenal Cortical-Medullary Circadian Rhythms." *Aviation, Space, and Environmental Medicine,* October 1976.

Siegel, V.; Gerathewohl, Siegfried J.; and Mohler, Stanley R. "Time Zone Effects." *Science,* June 1969, pp. 1249–1253.

Simpson, H. W. "A New Perspective: Chronobiochemistry." *Essays in Medical Biochemistry* 2 (1976): 115–181.

Simpson, H. W., and Halberg, F. "Phase Modulation of Desynchronized Human Circadian Adrenocortical Cycle on 21-Hr Day Simulating Circumglobal Travel (45° Longitude/Day)." *Endocrinology,* International Congress Series, 1972.

Simpson, H. W., and Stoney, P. J. "A Circadian Variation of Melphalan Toxicity to Murine Bone Marrow: Relevance to Cancer Treatment Protocols." *British Journal of Haematology* 35 (March 1977).

Smolensky, Michael S. "Rationale for Circadian-System Phased Glucocorticoid Management." In *Chronobiology,* edited by L. E. Scheving, F. Halberg, and J. E. Pauly. Tokyo: Igaku Shoin Ltd., 1974.

Smolensky, Michael; Halberg, Franz; and Sargent, Frederick, II. "Chronobiology of the Life Sequence." In *Advances in Climatic Physiology,* edited by S. Ito, K. Ogata, and H. Yoshimura. Tokyo: Igaku Shoin Ltd., 1972.

Smolensky, M. H.; Hartsock, P.; Lu, M. C.; Stevens, S.; Lagoguey, M.; and Reinberg, A. "Human Circadian Rhythms of Birth and Reproductive Function." Unpublished paper.

Smolensky, Michael H.; Tatar, Sheryl E.; Bergman, Stuart A.; Losman, Jacques G.; Barnard, Christian N.; Dacso, Clifford C.; and Kraft, Irving A. "Circadian Rhythmic Aspects of Human Cardiovascular Function: A Review by Chronobiologic Statistical Methods." *Chronobiologia* 3 (October–December 1976).

Stetson, Milton H., and Watson-Whitmyre, Marcia. "Nucleus Suprachiasmaticus: The Biological Clock in the Hamster?" *Science,* January 16, 1976.

Stroebel, Charles F. "Chronopsychophysiology." In *Comprehensive Textbook of Psychiatry,* vol. 2, edited by Alfred M. Freedman,

Harold I. Kaplan, and Benjamin J. Sadock. Baltimore: The Williams and Wilkins Company, 1975.

Strughold, Hubertus. "The Rhythmostat in the Human Body." Speech before the Texas Medical Association, October 24, 1974.

Strughold, Hubertus. *Your Body Clock.* New York: Charles Scribner's Sons, 1971.

Stumpf, Walter E., and Sar, Madhabananda. "The Heart: A Target Organ for Estradiol." *Science,* April 15, 1977.

Sturtevant, F. M.; Sturtevant, R. P.; Scheving, L. E.; and Pauly, J. E. "Chronopharmacokinetics of Ethanol." *Archives of Pharmacology* 293 (1976): 203–208.

Sulzman, Frank M.; Fuller, Charles A.; and Moore-Ede, Martin C. "Environmental Synchronizers of Squirrel Monkey Circadian Rhythms." *Journal of Applied Physiology* 43, no. 5 (1977).

Sulzman, Frank M.; Fuller, Charles A.; and Moore-Ede, Martin C. "Feeding Time Synchronizes Primate Circadian Rhythms." *Physiology and Behavior* 18 (1977).

Tasto, Donald L.; Colligan, Michael J.; Skjei, Eric W.; and Polly, Susan J. *Health Consequences of Shift Work.* Cincinnati: National Institute for Occupational Safety and Health, Robert A. Taft Laboratories, 1978.

Thomas, Lewis. *The Lives of a Cell.* New York: The Viking Press, 1974.

Timnick, Lois. "Body Rhythms May Be Linked to Illness." *Los Angeles Times,* October 10, 1978.

Underwood, Herbert. "Circadian Organization in Lizards. Role of the Pineal Organ." *Science,* February 11, 1977.

Ward, Ritchie R. *The Living Clocks.* New York: Alfred A. Knopf, 1971.

Weitzman, Elliot D. "Biologic Rhythms and Hormone Secretion Patterns." *Hospital Practice,* August 1976.

Weitzman, Elliot D. "Effect of Sleep-Wake Cycle Shifts on Sleep and Neuroendocrine Function." In *Behavior and Brain Electrical Activity,* edited by N. Burch and H. L. Altschuler. New York: Plenum Publishing Corporation, 1975.

Weitzman, Elliot D. "Neuro-endocrine Pattern of Secretion During the Sleep-Wake Cycle of Man." In *Progress in Brain Research,* vol. 42: *Hormones, Homeostasis and the Brain,* edited by

W. H. Gispen, Tj. B. van Wimersma Greidanus, B. Bohus, and D. de Wied. Amsterdam: Elsevier Scientific Publishing Company, 1975.

Weitzman, E. D. "Temporal Organization of Neuroendocrine Function in Relation to the Sleep-Waking Cycle in Man." In *Recent Studies of Hypothalmic Function.* New York–Basel: S. Karger, 1974.

Weitzman, Elliot D.; Fukushima, David; Nogeire, Christopher; Roffwarg, Howard; Gallagher, T. F.; and Hellman, Leon. "Twenty-four Hour Pattern of the Episodic Secretion of Cortisol in Normal Subjects." *Journal of Clinical Endocrinology and Metabolism* 33, no. 1 (July 1971): 14–22.

Weitzman, Elliot D.; Kripke, Daniel F.; Goldmacher, Donald; McGregor, Peter; and Nogeire, Chris. "Acute Reversal of the Sleep-Waking Cycle in Man." *Archives of Neurology,* June 1970.

Weitzman, Elliot D.; Nogeire, Christopher; Perlow, Mark; Fukushima, David; Sassin, Jon; McGregor, Peter; Gallagher, T. F.; and Hellman, Leon. "Effects of a Prolonged 3-Hour Sleep-Wake Cycle on Sleep Stages, Plasma Cortisol, Growth Hormone and Body Temperature in Man." *Journal of Clinical Endocrinology and Metabolism,* June 1974.

Weitzman, Elliot D.; Perlow, Mark; Sassin, Jon F.; Fukushima, David; Burack, Bernard; and Hellman, Leon. "Persistence of the Twenty-four Hour Pattern of Episodic Cortisol Secretion and Growth Hormone Release in Blind Subjects." *Transactions of the American Neurological Association, 1972.* American Neurological Association. New York: Springer, 1973.

Wever, Rütger. "The Circadian Multi-oscillator System of Man." *International Journal of Chronobiology* 3 (1976): 19–55.

Winget, Charles M.; Hughes, Lewis; and LaDou, Joseph. "Physiological Effects of Rotational Work Shifting: A Literature Survey." *Journal of Occupational Medicine* 20, no. 3 (March 1978).

Wurtman, R. J. "The Effects of Light on the Human Body." *Scientific American,* July 1975.

Wurtman, Richard J. "Effects of Light and Visual Stimuli on Endocrine Function." *Neuroendocrinology* 2 (1967).

Wurtman, Richard J., and Moskowitz, Michael A. "The Pineal Organ." *New England Journal of Medicine* 296 (June 9 and 16, 1977).
Zacharias, Leona; Rand, William M.; and Wurtman, Richard J. "A Prospective Study of Sexual Development and Growth in American Girls: The Statistics of Menarche." In *Obstetrical and Gynecological Survey.* Baltimore: The Williams and Wilkins Co., 1976.

Seminars

Workshop on "Biological Clocks." Hopkins Marine Station, Stanford University, Pacific Grove, Calif. June 1977.
International Symposium on Clinical Chronopharmacy, Chronopharmacology, and Chronotherapeutics. School of Pharmacy, Florida A & M University, Tallahassee, Fla. February 1978.
"Concepts of Stress." The University of Southern California, Los Angeles, Calif. March 1978.
Association for the Psychophysiological Study of Sleep. Stanford University, Palo Alto, Calif. April 1978.